# The Internet as Second Ac

One of the most significant and important advancements in information and communication technology over the past 20 years is the introduction and expansion of the Internet. Now almost universally available, the Internet brings us email, global voice and video communications, research repositories, reference libraries, and almost unlimited opportunities for daily activities. Bridging geographical distances in unprecedented ways, the Internet has impacted all aspects of our daily lives – facilitating the running of businesses, the attainment of services and keeping in touch with friends and family. Accessible at any time and for many of us from our mobile phones, the Internet has opened up a world of knowledge and communication platforms that we cannot now imagine living without. This book explores the concept that the Internet has become a second action space for individuals. Coexisting with traditional and "obvious" real space, the Internet serves as a novel spatial platform and action space to its subscribers all over the world. Kellerman expertly discusses this notion and examines the practical integration of cyberspace with real space.

Part I examines the Internet as a platform for action and presents its relations with physical space concerning a range of uses and applications which were traditionally performed in physical space only. It discusses the idea that the Internet has become a second space and explores theoretical perspectives surrounding this notion. The Internet has undeniably made humankind more efficient and connected. Part II explores the Internet as an action space for human life, considering basic human needs, curiosity, identity and social relations. It further considers instances whereby use and application of the Internet cannot be fully performed in real space, mainly regarding people's presentation of identity. Part III explores daily actions over the Internet, such as work, shopping, banking and social interactions. Kellerman also briefly touches on the darker aspects that the expansion of the Internet has made possible – including its role in fraud and other crimes. The concluding chapter discusses people living across the two spaces and identifies potential future developments.

*The Internet as Second Action Space* will appeal to students across the social sciences, in particular those studying Geography, Sociology, Media Studies, Internet Studies, Business and related disciplines.

**Aharon Kellerman** is a Professor Emeritus in the Department of Geography, University of Haifa, Israel, and President, Zefat Academic College, Israel. He also serves as Vice-President of the International Geographical Union (IGU), and acts as Honorary Chair of its Commission on the Geography of the Information Society, which he established and chaired. He has published six books; five monographs; over 70 refereed articles; 60 book chapters; numerous conference proceedings and book reviews.

As cyberspace and physical places become ever more enfolded into one another, the Internet has become an increasingly important arena through which social relations and individual identities play out. Aharon Kellerman has long been an astute observer of these trends. This volume artfully charts how many activities once confined to the physical domain have moved steadily online, with surprising and often unintended consequences. Increasingly, many human needs, from the most basic to the most sophisticated, have shifted to the Web, changing identities and interpersonal relations as a result. This volume lucidly depicts the process through which cyberspace and real space act both as complements and as substitutes in multiple domains of daily life. Kellerman avoids the glossy utopianism that pervades many popular depictions of the Iternet to reveal its "dark side," the world of gambling, pornography and hacking that inevitably accompany the gains in productivity that it unleashes.

Professor Barney Warf, *University of Kansas, US*

Aharon Kellerman continues his exploration of cyberspace in The Internet as Second Action Space, his latest expedition and third major book on the geography of our Internet-enabled and mediated world. The book provides a valuable interpretation of human behavior when offered new places and spaces to interact by applying the rigor of geography to a rapidly growing and changing phenomenon. By mapping and framing our use of cyberspace, Kellerman provides a text essential to our understanding of virtual action spaces.

Professor Mark Wilson, *Michigan State University, US*

# The Internet as Second Action Space

**Aharon Kellerman**

**Routledge**
Taylor & Francis Group

LONDON AND NEW YORK

First published 2014 by Routledge

2 Park Square, Milton Park, Abingdon, Oxfordshire OX14 4RN
711 Third Avenue, New York, NY 10017

*Routledge is an imprint of the Taylor & Francis Group, an informa business*

First issued in paperback 2018

*British Library Cataloguing in Publication Data*
A catalogue record for this book is available from the British Library

*Library of Congress Cataloging in Publication Data*
Kellerman, Aharon.
The Internet as second action space/Aharon Kellerman.
    pages cm
    1. Internet–Social aspects. 2. Cyberspace–Social aspects. I. Title.
    HM851.K446 2014
    302.23′1–dc23                                    2014003121

ISBN: 978-0-415-85871-7 (hbk)
ISBN: 978-1-138-37793-6 (pbk)

Typeset in Times New Roman
by Wearset Ltd, Boldon, Tyne and Wear

**Dedicated to my granddaughter Shani-Naomi**

# Contents

x   *Contents*

# Figures

# Tables

# Preface

This book attempts to explore the contemporary daily and routine experiences of individuals, mostly in developed countries, who have positioned themselves and who function within a new arena for human action, that of virtual space. This virtual space of human action has emerged mainly over the Internet, through which it also operates. The contemporary performance of individuals in virtual space has made them function in parallel in two spaces: the real and the virtual. Kinsley (2013, 4–5) has recently argued on the study of life in virtual space that "concerted studies of the evolution and increasing significance of the digital in our lives have waned." This book attempts in some way to fill this lacuna. The book will, therefore, highlight the new virtual space as an action space for individuals, and we are, thus, about to explore in the following chapters the origins of this virtual action space and its relations with real space, and most importantly the rules and the ways and means for its operation in numerous spheres.

For me personally as a researcher, this book on the Internet as second action space constitutes an additional step of my study ladder of virtual space on which I have attempted to climb during the last three decades. This ladder consists of the three components of telecommunications, the Internet and personal mobilities. For all of these three components, which comprise both means and technologies for human action I have attempted at the time to explore their very innovation and development, side-by-side with the study of their societal adoption and their socio-spatial dimensions. I believe that the ability to portray presently, in this book, a widely functioning virtual space for human action, a space that is now available for us through smartphones literally wherever we go within physical space, implies a maturing of the Internet in particular, and of virtual space in general, from the perspective of human users. Thus, in some way this book constitutes a continuation to two of my previous works: *The Internet on Earth: A Geography of Information* (2002), and *Daily Spatial Mobilities: Physical and Virtual* (2012). The first of these two books portrayed a down-to-earth geography of the Internet, and now, slightly a decade later, the Internet has already gained its own virtual geography, as presented in this book. The second book presented virtual mobilities as a family of mobility media emerging next to physical mobility media, and this book will attempt to explore the destination of much of these virtual mobilities, namely virtual space per se.

The very idea that the computerized world is second to the real one in some ways is basically not a new one. Back in 1985, almost a decade before the emergence of the Internet in its current form, Turkle authored a book entitled *The Second Self: Computers and the Human Spirit*. The contemporary Internet actually complements computers as second selves, in its constitution as a second action space, alongside with real space. Furthermore, the very basic recognition of the Internet as a second action space probably dates back to 2008, when Yu and Shaw stated rather generally though that "with both physical and virtual spaces available to us today, individual opportunities for potential activities can exist not only at locations within the individual's physical proximity, but also wherever they can be reached through tele-presence" (p. 410).

I wish to express my gratitude to all those who assisted my writing, directly and indirectly. Thanks are due to all commentators on my paper presentations in numerous conferences over the last few years, notably in the recent Kyoto 2013 Regional Conference of the International Geographical Union (IGU), as well as in the Montreal 2013 World Social Science Forum (WSSF) of the International Social Science Council (ISSC). Thanks are also due to Noga Yoselevich (University of Haifa) and to Moshe Asaf (Zefat Academic College) for the processing of Figure 3.1, and to Kety Gersht (Zefat Academic College) for data preparation for the Appendix, as well as for the integration of the figures with their titles. The Research Authority of the University of Haifa assisted the completion of this book through the funding of index preparation. As always, I owe a great debt to my wife, Michal, whose sharing and patience have made the conception and the writing of this book possible.

December 2013

# Acknowledgments

The following previous publications of the author were used for the writing of this book: *The Internet on Earth: A Geography of Information* (London: Wiley, 2002); *Personal Mobilities* (London and New York: Routledge, 2006); *Daily Spatial Mobilities: Physical and Virtual* (Farnham and Burlington, VT: Ashgate, 2012); M.I. Wilson, A. Kellerman and K.E. Corey, *Global Information Society: Knowledge, Mobility and Technology* (Lanham, MD: Rowman and Littlefield, 2013); "The satisfaction of human needs in physical and virtual spaces," *The Professional Geographer*, DOI: 10.1080/00330124.2013.848760.

# Part I

# The Internet as a platform for action space

# 1 The Internet as second space

The geography of the Web is as ephemeral as human interest.

(Weinberger 2002: 6)

## Opening background

Our discussions in this book will be based, at least implicitly, on the notion that the Internet "links millions of people in new spaces that are changing the way we think, the nature of our sexuality, the form of our communities, our very identities" (Turkle 1995: 9). As such, the Internet has been variously assessed. By its contents, some of which we will review during the course of the book, the Internet has been viewed "as a library, a telephone, a public park, a local bar, a shopping mall, a broadcast medium, a print medium, a medical clinic, a private living room, and a public educational institution" (Biegel 2001: 28). The very human expansion into virtual space has brought some scholars to believe that "this new round of spatial expansion could have far greater economic and social implications than the discovery of America by Columbus. Columbus discovered only one new continent, but many new virtual worlds are being created" (Li, Papagiannidis and Bourlakis 2010: 426). The following discussions should provide at least some food for thought in this direction.

The Internet, as a computerized textual communications system, was originally trialed in the US in the 1960s in the form of an electronic mail network. It was originally developed in order to serve as an alternative security network for the telephone and telex systems in case of a nuclear attack. Its current universal availability is an excellent example for the adoption of a technology for purposes completely different than those envisaged by its developers. Since the mid-1990s the Internet has been stabilized as including two components: a universal and global email system, namely as a communications system, side-by-side with the World Wide Web (WWW), which constitutes the largest library and information storage entity worldwide, dynamic and interactive, thus constituting an information system. The integration of these two Internet components of communications and information into one system has permitted instant and worldwide availability to individuals of both personal and public information, as well as their own contribution to both types of information. By the end of 2011, one-third of the global population enjoyed access to the system (70 percent in developed countries) (ITU 2012).

The Internet enjoys numerous features and characteristics. It is a free and an uncontrolled medium for personal virtual mobility, so that there are no institutional or governmental laws and regulations for its "driving" or for "passing through" its communications lines. It further provides for co-presence of individuals in both physical and virtual spaces simultaneously. Emailing permits long distance and international messages and calls at no charge per call or by time, and the Web expands the lived-spaces of its users beyond their real locations, since cyberspace represents an additional lived space, as we will see in the following sections of this chapter, as well as in later chapters.

Socially, the Internet has expanded human rights by its very provision of instant written communications, as well as through its provision of access to information. Practically, it has extended people's virtual personal and public expressions to unprecedented levels, through both the Web, notably through social networking over Web 2.0, and obviously through emailing. One basic dimension of these extended mobility and speech expressions have been the emerging virtual communities and networks, which we will discuss in Chapter 8.

The most dramatic trend in information and communications technology (ICT) innovations over the recent years has been the effort to bring about the maturity of the Internet making it become fully mobile and universally available, thus permitting access to both email and the Web at any time and in any place, through wireless fidelity (Wi-Fi) and third generation (3G) cellular modems, both available mainly through smartphones, as well as through laptops and tablets. This universal and instant availability of the Internet in real space has implied a practical integration of cyberspace with real space, as we will see in this and the following chapters (see also Kellerman 2010).

Mobile telephone technology was originally introduced in 1906 by Lee de Forest. The first and rather limited mobile telephone services were introduced in the UK in 1940 and in the US in 1947, followed by the commercial introduction of such services in 1979–1983 through the allocation of wavelengths for its operation (Kellerman 2012a). Since then, the mobile phone has turned out to be the most widely diffused communications device globally, with a global penetration rate of 86 percent by the end of 2011. Some 90 percent of the world population was covered by a mobile phone signal already in 2009. About 14.3 percent of the global population, mainly in developed countries, enjoyed access to mobile broadband services by the end of 2011, with an annual growth rate of this access of 40 percent! (ITU 2012).

The smartphone, originally introduced in 1993 and diffusing widely as of the 2000s, has actually been developed into an advanced laptop, which includes also additional features that used to be available only through dedicated devices, such as Global Positioning System (GPS) for road navigation, and high-quality cameras. Another smartphone feature is a new generation of Location Based Services (LBS), providing commercial local information (e.g., on restaurants) into people's mobile phones by their locations. Furthermore, the availability of mobile broadband for connection to the Internet has implied the more extensive and more instant uses of Internet applications developed originally for fixed broadband communications. Striking such uses are for entertainment purposes, permitting radio and cellvision reception, as well as the viewing of streaming pictures in form of video clips and full movies. In addition, several other applications became possible, notably e-work (synonymously called telecommuting or telework), e-banking, e-government, e-health, e-learning (synonymously called distance learning), and business to customers (B2C) e-commerce (synonymously termed online shopping). These will be highlighted and discussed in Chapter 7. At the same time, Short Message Services (SMS) have turned from a mere inter-personal communications medium into a business-to-clients communications channel as well.

## Book objectives and structure

The discussions in this and in the following chapters aim at the presentation, analysis and interpretation of the Internet as a rider on cyberspace, thus turning the latter into a second action space for contemporary individuals, mainly in developed countries, side-by-side with their traditional and "obvious" action spaces located in real space. Obviously the rather selected mesh of uses and applications pursued by individuals over the Internet which we will discuss in the following chapters is not comprehensive, given the enormously wide array of uses and applications actually performed over the Internet, but its wide range should provide a reasonable insight into the actions performed by individuals over the Internet. In addition, we will discuss in this book applications and uses which could not be fully performed in real space, before the maturing of the Internet, mainly regarding people's identity presentation (Chapter 6).

The book is divided into three parts: the Internet as a platform for action space; human needs and the Internet; and the Internet as an action space for individuals. The first part includes three chapters: an introductory chapter on the Internet as second space (Chapter 1), which will lead us to an elaboration of theoretical perspectives on the Internet as second action space (Chapter 2); and finally we will get acquainted with Internet operations and economies (Chapter 3). The second part also consists of three chapters, the first of which will focus on human basic needs and their provision (Chapter 4), followed by expositions of curiosity and its satiation (Chapter 5); and personal identity over the Internet (Chapter 6). The third part of the book consists of four chapters, beginning with elaborations on daily activities (Chapter 7); and continuing with discussions on social relations (Chapter 8), as well as on darker actions over the Internet (Chapter 9). The book will then conclude with a chapter focusing on the time and space of people living in both real and virtual two spaces as well as on the potential future action status of the two spaces (Chapter 10).

The rest of this chapter is devoted to several introductory perspectives on the Internet as a second space. First, we will present cyberspace, focusing on its nature, classes and cognition, followed by a comparison between real and virtual spaces, and a discussion on the practical relations between real and virtual spaces. We will then introduce the Internet, its history and characteristics, followed by an exposition of mobile communications technologies. Next, some comments on the auditory geography of the Internet as compared to that of real space will be elaborated on. These discussions will be followed by a presentation of the notion of "action space," notably with reference to cyberspace and the Internet.

## Cyberspace: nature, classes and cognition

The introduction of the Internet in its current form as of the mid-1990s has brought about a focus on *cyberspace*, a term which was originally proposed by Gibson (1984) as a science-fiction notion, and which was applied later to computer-mediated communications as well as to virtual reality technologies (Kitchin 1998: 2). Hence, since the early 1990s cyberspace has been variously defined as:

1   *Artificial reality*: "Cyberspace is a globally networked, computer-sustained, computer-accessed, and computer-generated, multidimensional, artificial, or 'virtual', reality" (Benedikt 1991: 122, see also Kitchin 1998: 2).
2   *Interactivity space*: "interactivity between remote computers defines cyberspace ... cyberspace is not necessarily imagined space – it is real enough in that it is the space set up by those who use remote computers to communicate" (Batty 1997: 343–4)
3   *Conceptual space*: "the *conceptual space* within ICTs (information and communication technologies), rather than the technology itself" (Dodge and Kitchin 2001: 1).

These three definitions may be seen as complementary rather than contradictory to each other, so that we can view cyberspace as constituting simultaneously a virtual, interactive and conceptual entity. The common thread among these three definitions of cyberspace is that cyberspace has been viewed as a category of space or reality. In addition, cyberspace may represent real space through maps, pictures and graphs, which may then be used for an understanding of real space and for navigating in it (Zook and Graham 2007a). Further in this regard, Larsen *et al.* (2006) proposed a differentiation among *imaginative travel* through images and memories, *virtual travel* through the Internet and *communicative travel* via letters, phone calls and emails. However, this differentiation lost much of its practical significance, given the growing convergence among mobile phones, the Internet and TV.

Cyberspace may be divided into two classes or components in terms of purposes and uses: information [cyber]space (mainly the Web) and communications [cyber]space (mainly email and Web 2.0 applications, such as Facebook and Twitter) (Kellerman 2007). Information space refers to digital information sets or systems, consisting of information organized within spatial contexts such as websites, and, hence, involving geographical metaphors such as sites, homes and navigation/surfing. Information cyberspace further refers in general to digital information sets, such as data archives and library catalogs (Fabrikant and Buttenfield 2001; Couclelis 1998). All these information sets are textual and/or graphic in nature and they have some constancy in terms of their virtual availability to users, so that they may be recalled whenever users find it necessary. Most of these information files are meant to be shared by users: either the general public through the Internet, or segmented and permitted users only, through Intranets. Contemporary search engines have allowed for easy access to websites and files. Google has emerged as a leading service in this regard, providing also searches into specialized information systems, such as satellite images and scientific articles and books, thus turning Google into what one termed as a megaproject within another megaproject (the Internet) (Paradiso 2011, see also Chapter 3).

The second class of cyberspace is communications cyberspace, referring to the cyberspace of persons who communicate with each other via numerous modes of communications: first, and foremost, through video communications, which includes also the transmission of real space visible in the background of the communicating parties, as well as the transmission of the images of the callers themselves; and second, through purely electronic and invisible spaces of interaction, using non-video communications media (mainly written emails, faxes, SMSs, chat platforms and audio telephone calls) (Kellerman 2007). Communications [cyber] space is mostly interpersonal or shared by small groups, though it may be more widely accessible to larger groups through social networking systems, such as widely distributed blogs, or through networking platform, such as MySpace, Facebook, Twitter and Usenets (see Chapter 8). Much of the contents of communications cyberspace is not recorded, and if recorded the contents is meant to be shared only by the communicating parties.

The two categories of digital/virtual spaces of information and communications are frequently interlinked, for example when emails are sent through an informative website rather than through an email interface, or in emails and messages transmitted through Usenets, blogs, Facebook and MySpace, which include links to pictures, websites, and/or data. Such interfolded and even fused information and communications [cyber]spaces attest to the oneness of the Internet from the usage perspective, thus extending much beyond the shared telecommunications infrastructure of the two spaces. However, each of the two cyberspace classes may frequently function independently of the other, for instance oral personal communications normally does not involve a simultaneous transmission of textual datasets.

The very use of cyberspace has involved, until the introduction of broadband for desktop PCs and mobile broadband for laptops and smartphones, some preparations for the very access of computers and their connection to the Internet, since charging for Internet use was by the duration of use sessions. Following the introduction of broadband in the late-1990s computers and mobile phones have become continuously connected to the Internet by broadband subscriptions, as the charge for its use was no longer determined by use duration.

The use of the Internet from its outset has been typified by co-presence of users in both fixed physical location and in virtual spaces (see, e.g., Kaufmann 2002: 28; Urry 2000: 71), something which others preferred to as simultaneous embodied and response presences (Knorr-Cetina and Bruegger 2002). Adams (1995) argued that in the pre-Internet era human extensibility constituted a "transcendence of place" (p. 269), and the media amounted to "extensions of man" (p. 269), notably at times of growing globalization. However, the introduction of mobile Internet connection through Wi-Fi and cellular modems has turned the exposure to the Internet into an instant and rather permanent condition, first through laptops and later, as of the early 2000s, also through mobile phones, notably through the so-called smartphones. The mobile nature of the use of smartphones has made it more difficult to clearly identify co-presence, or simultaneous embodied and response presences, notably from the perspective of users. Furthermore, Arminen (2007) noted "a dual nature for mobile media, making them both global and local" (p. 432), permitting distant presentation of the self, thus blurring the difference between the social nature of global and local contacts (Knorr-Cetina and Bruegger 2002).

Internet users may over time comprehend and get used to surfing procedures for reaching selected websites or software and then navigate through them, but they may have difficulties to cognize and eventually draw cognitive maps of virtual landscapes or of virtual cartographic maps which they have been exposed to through rather restricted sensory viewing over the Internet. Furthermore, virtual landscapes or maps can be manipulated in diversified ways, such as by changing their scale, size, directions, colors, richness of information, and so on. By the same token, virtual texts can also be manipulated through changing of their formats, fonts, color and so on. Such manipulations may add to the difficulty of cognizing cyberspace presentations in memorable ways. As Kwan (2001) noted for real space, space and its maps are two completely separated entities, whereas in cyberspace they may converge.

When communicating over the telephone, parties may cognize a metaphorical space that merely permits a feeling of "presence" and intimacy by the communicating parties:

> "The virtual" is imagined as a "space" between participants, a computer-generated common ground which is neither actual in its location or coordinates, nor is it merely a conceptual abstraction, for it may be experienced "as if" lived for given purposes.
>
> (Shields 2003: 49)

Cognitive communications cyberspace constitutes a component of interpersonal communications among people, but the contextual elements of each cognitive entity may be different, since cognitive space is dominated by changing physical elements, such as paths, landmarks, etc., whereas people are not necessarily part of such cognitive space. However, this spatial context lies in the shadow of people, the communicating parties, who dominate the cognitive communications cyberspace, so that the surrounding physical environment, which may be viewed in video conversations, constitutes background only. Alongside the communicating parties and the spatial context surrounding them, there are in each call several additional elements, pertaining to human contacts. Language, by its very nature, is essential for communications notably for online spoken or written communications, which does not leave time for translation, something which is possible in the use of data or information in information space. Furthermore, the time framework is also essential if communications takes place in real time, and when the two parties are located in different domestic or international time zones. Such time differences apply not only to daily differences but also to weekend extents and holiday differences among nations, as well. Less crucial additional elements are weather conditions, which may potentially deter communications, and international currency differences, if merchandise and services are sold/bought internationally. Altogether, the quality of communications is of much importance, notably for video conversations. Such video conversations may profit from broadband transmissions probably more than other forms of communications and information transmissions.

Following Shum's (1990) distinction between locational and attributional information included in traditional cognitive maps, we may claim that the cyberspace of interpersonal communications, as well as the cognitive cyberspace for such communications, include attributional information with no, or just little, locational information, as we have noted above. Furthermore, cognitive communications cyberspaces are personally unique, and cannot be aggregated, whereas cognitive maps relating to a specific area may be compared and conclusions on a wider societal knowledge of an area drawn. Cognitive space and cognitive mapping may facilitate *spatial behavior*, or corporeal personal mobility, whereas cognitive communications cyberspace may facilitate *social and economic behavior*, in form of interpersonal communications, performed following a specific call, or some online shopping activity following a more commercial interaction, respectively. Related to this, cognitive space/maps may facilitate navigation or movement *in* places, whereas cognitive communications cyberspace may facilitate movement *to* other places following virtual contacts made with people in them.

## Comparing real and virtual spaces

The very extension of the notion of space from the material world to the virtual one has called for explorations of possible practical relations between the two classes of real and virtual spaces. Shields (2003: xv) referred in this regard to a contemporary "shifting relationship between the virtually real, and the material." Moreover, Crang *et al.* (1999) argued for "the virtual as spatial" (p. 11), and that "virtuality [then] is not just something which operates through and across space. It is at its heart a spatial phenomenon" (pp. 12–13). Thus, cyberspace was recognized to be "hardly immaterial in that it is very much an embodied space" (Dodge 2001: 1). The recent introduction of broadband connectivity, which has permitted permanent access to cyberspace, has turned these observations and assessments even more meaningful, in that cyberspace has become completely integrated into daily activities performed in physical space. Already before the introduction of broadband, cyberspace was interpreted by Bolter and Grusin (1999: 179) as a virtual form of Augé's (2000) *non-places*, which he originally proposed for real, but unpopulated, spaces, such as airports.

Cyberspace may be viewed also as being embedded in physical space: "electronic space is embedded in, and often intertwines with, the physical space and place" (Li *et al.* 2001: 701). Furthermore, both spaces, the physical and the virtual, co-evolve in that they "stand in a state of *recursive interaction*, shaping *each other* in complex ways" (Graham 1998: 174). The power of cyberspace in its constitution of a virtual entity, which presents and represents real space will be highlighted in the next chapter devoted to theoretical dimensions of virtual space. However, it is important to comment already here that the Web, as an application of, or as a rider on the rather wider entity of cyberspace, may be viewed as a special form of social space. Like the more real social space it constitutes a resource and a production force, for instance for its provision for online shopping. Further like real social space, Web applications in cyberspace may be looked upon, by their very nature, as texts and as symbols for both individuals and organizations, and they may further serve as organizational frameworks notably in Intranet systems within work places and multi-location companies (Kellerman 2002). Several commentators claimed for cyberspace per se to constitute also a landscape, a place and even a social value (see Dodge and Kitchin 2001 for detailed discussions), and we will return to this topic again in the next chapter. Such views substantiate the third definition of cyberspace as conceptual space, which we noted in the previous section. Needless to say, though, that in its Web component, cyberspace constitutes an *imagined* space of representation, through its virtual imitation or virtual description of real spaces and places.

Real space and cyberspace have also been viewed as interfolded into each other: cyberspace is accessed from real space, and it further contains data on material space, and thus affects it. This relationship has become powerful with the introduction of images of real space by Google through Google Maps and Google Earth for satellite images, and Google Street for real space pictures, permitting users to manipulate the Google produced materials into their own images and to use them for spatial navigation, thus creating DigiPlaces, or blends of the digital and the real (Zook and Graham 2007a, 2007b, 2007c; Crutcher and Zook 2009). Furthermore, cyberspace keeps or imitates several features of physical space. "Virtual environments contain much of the essential spatial information that is utilized by people in real environments" (Péruch *et al.* 2000: 115). However, cyberspace was argued to have its own geography and to be symbol-sustained (Benedikt 1991: 123, 191; Batty 1997). It was further suggested that cyberspace enables and constraints its users similarly to real space (Adams and Ghose 2003), and it was further argued that digital space has its own materiality (Kinsley 2013).

Side-by-side with the connections and convergences between real and virtual spaces, cyberspace is distinguished from real space in many instances (Table 1.1). These differentiations may be divided into three groups of dimensions:

Table 1.1 Real and virtual spaces

| Dimension | Real space | Virtual space |
|---|---|---|
| *Organization* | | |
| 1 Content | Physical and informational | Informational |
| 2 Places | Separated | Converge with local real ones |
| 3 Form | Abstract or real | Relational |
| 4 Size | Limited | Unlimited |
| 5 Construction and maintenance | Expensive and heavily controlled | Reasonably priced and lightly controlled |
| 6 Space | Territory/Euclidean | Network/logical |
| 7 Matter | Material/tangible | Immaterial/intangible |
| *Movement* | | |
| 8 Medium | Transportation | Telecommunications |
| 9 Speed | Depends on the mode of transportation | Speed of light, constrained by infrastructure, costs, regulations etc. |
| 10 Distance | Major constraint | Does not matter mostly |
| 11 Time | Matters | Matters, but events can suspend in time |
| 12 Orientation | Matters | Does not matter |
| *Users* | | |
| 13 Identity | Defined | Can be independent of identity in real space |
| 14 Experience | Bodily | Imaginative, metaphorical, close to reality |
| 15 Interaction | Embodied | Disembodied |
| 16 Attitude | Long-term commitment | Can also be uncommitted |
| 17 Language | National-domestic | Mainly English-international |

Sources: Table: Kellerman 2002: 35; Kellerman 2007; Kellerman 2010b; Kellerman 2012a: 119, Items 1–2, 8–12: Li *et al.* 2001; items 3, 14–15: Dodge and Kitchin 2001: 30, 53; items 6–7: Graham 1998.

organization, movement and users, and they are discussed in detail elsewhere (see Kellerman 2002: 33–8 and Kellerman 2012a: 119). It will suffice here to note that the physical space of virtual space, namely the fixed computer hardware required for its operation, may be viewed from a spatial perspective as an auxiliary physical space for accessing cyberspace by users. This physical space/ hardware seems to have lost its spatial significance once access to virtual space has become fully mobile through mobile phones and portable computers.

Virtual and real spaces are also interrelated in more indirect ways. For instance, it was shown that persons who navigated successfully through a virtual maze presented also a more successful way-finding in real space (Péruch *et al.* 2000). Generally, then, information, knowledge and experience constitute mediating forces between the construction and the reshaping of both real and virtual spaces.

From a cultural-geographic perspective, cyberspace was argued to amount to a kind of a Heavenly New Jerusalem (Benedikt 1991; Wertheim 1999). As such, cyberspace was assumed to move contemporary society from a mere conception of physical space into a more complex one involving also an inner spiritual space, achieved through social networks which are based on global information sharing (Wertheim 1999).

The discussion in this section has attempted to show that cyberspace is simultaneously an entity of its own as well as an entity converged with real space. The convergence of real and virtual spaces applies mainly to services which are offered in complementary forms in the two spaces, such as shopping and banking, as well as to cyberspace as representing real space, through services such as Google Maps, Google Street and Google Earth. The recently growing tendency of cyberspace to converge with real space has appeared through the removal of the distance or the barriers that separated between these two classes of space, in the form of cumbersome and costly Internet connectivity and in the form of slow and restricted pace and volume of activity. The distance and barriers have been removed with the introduction of broadband services and even more so with the emergence of mobile broadband services. The instant access to cyberspace offered by broadband services has made users take for granted the access to the Internet and the use of its services and, hence, see the Internet as fully integrated into their daily routines. This applies even stronger for Web 2.0 social networking applications, introduced as of the early 2000s, which call for continuous or very frequent attention by their users, permitted mainly by mobile broadband connectivity, notably through hand-held smartphones.

## Practical relations between real and virtual spaces

Real social space is not independent of virtual space anymore. It could be argued, until the introduction of information technology, that the very constitution of material space was relatively independent of imagined ones, followed later by some complementarity between real space and early introductions of cyberspace (see Lee *et al.* 2001). It seems that the tremendous integration of information technology and Internet into all spheres of economic and social lives, makes it impossible now to manage and manipulate real social space without the use of, or reference to, virtual space. This may relate, for instance, to smart buildings, the utilities and systems of which are operated by a unified computerized system, replacing conventional ones (see Batty 1997). It may further apply to the maintenance of conventional real estate as well using separate software and hardware for each utility. By the same token, the production and consumption of material products and services in real space are interwoven with Internet activities, whether it be for the obtaining of relevant information, for the performance of business correspondence, for maintenance assistance, or for consumption through cyberspace (e-commerce or online shopping). The production and consumption of non-electronic information, such as books, is also interwoven with virtual space, through book editing, for example, at the production end, as well as for browsing and eventual book purchase at the consumption end.

Attempts to define the exact nature of the relationship between real and virtual spaces have been presented and commented on by numerous commentators. Thus, Batty (1997: 341) suggested that "space and place has been influenced by the gathering momentum of the digital world," and he further referred to "the impact of computers and communications on place itself." Other commentators have noted more complex patterns of interrelationship. Thus, "cyberspace is hardly immaterial in that it is very much an embodied space" (Dodge 2001: 1), and from the other end, "information systems redefine and do not eliminate geography," and "electronic space is embedded in, and often intertwines with, the physical space and place" (Li *et al.* 2001: 701). Thus, Internet "is shaped by, and reflects, the place-routed cultures in which it is produced and consumed" (Holloway and Valentine 2001: 153).

Graham (1998) went one step further than the previous formulations for absolute or relational views of real and virtual spaces, by suggesting that both spaces co-evolve in that they "stand in a state of *recursive interaction*, shaping *each other* in complex ways" (p. 174). This two-way process, may bring about a "liberation" of activities previously taking place in real places only, but their moving into cyberspace creates new real and fixed locations for the functioning of cyberspace (see Swyngedouw 1993). This may be the case, for example, for the location of huge warehouses for online stores, such as those of Amazon in the State of Washington (US).

Summing up the numerous observations so far, we identified several interelationships between real and virtual spaces:

1  *interdependence* in their very functioning;
2  *co-evolution* of both spaces;
3  *dual construction and elimination* of sites and activities in both spaces.

Information, the only content of cyberspace, is produced by people located in real space, but its very production and consumption are dependent and are embedded in virtual space, the major contemporary medium for its production, transmission and consumption. The construction of new websites may bring about the development of real world industries and services, and the other way around, as well, since the very existence of industries and services in real space, may bring about further developments in cyberspace (inform of additions to existing websites or the construction of new ones). At the same time, the availability of information services over cyberspace, handled by computers located in some specific centralized real world places, may bring about a partial elimination of local facilities (e.g., travel agencies).

## History and characteristics of the Internet

The Internet was originally invented in the US in 1969, as the ARPANET (Advanced Research Projects Agency Network) network, a network that constituted an experimental alternative communications system for telephone services, developed for a potential replacement of the telephone system in case of nuclear disasters. As such, it was originally experimented through a network connecting security headquarters with universities (Kellerman 2002), followed by the emergence of academic networks (e.g., BITNET [Because It's Time Network or Because It's There Network] and NSFNET [National Science Foundation Network]). It took a long period of some 25 years of incubation and development for these early security and academic electronic networks of communications and information until they matured into a universally open and commercial entity, the Internet, in 1994. However, it took much less time, just seven years following its introduction, in 2001, that the Internet was adopted by one half of Americans, either having access to it or being online at home. Its current universal availability has been considered as the best example for the adoption of a technology for purposes completely different than those envisaged by its developers (Urry 2003: 63).

Compared to the diffusion of the telephone, the rapid adoption of the Internet, as well as that of mobile telephony, has had to do with the prior existence of partial telecommunications infrastructures for its operation, through the telephone system, so that new connections to the system could be easily performed. Of no less importance, though, is the emergence of the Internet and mobile telephony at a time when these innovations constituted technologies and means which have supported the evolving information society, based on standalone computers as well as the digitization of the previously existing telephone system. The information society, on its part, has placed a special emphasis on the production, processing, transmission and consumption of information. This point is strongly demonstrated by the slow evolution of mobile telephony at the time of its original invention, early after the introduction of the telephone. Mobile telephony had rather to await its final development until the release of the required wave spectrum in the late 1960s, when proper social and economic conditions emerged for such a long-awaited move by the American FCC (Federal Communication Commission).

Personal telecommunications at large, and the Internet in particular, have been governed by an *open code*, which Lessig (2001) considered as the "heart of the Internet" (p. 246). This open code permits unlicensed access for the production of Internet information, whether through the establishment of websites or through the writing of email messages. It further permits an open access to the consumption of Internet information, through the receipt of emails, as well as through accessing free of charge websites. The open code principle further permits the uncontrolled flows of information from any origins to any destinations, unless controlled by governmental censorship (see Chapter 9). This open code system also allows for innovations to be freely introduced and adopted for both the production and consumption of Internet information. All these activities are restricted neither by a minimal nor by a maximal age of users, so that the use of the Internet constitutes a completely informal activity, as compared, for example, to the requirement for driving licensing.

This nature of the Internet may be related to its origin in the US and the accent of American society on freedom of expression. The open code nature of the Internet has been questioned by numerous forces of a regulatory nature, such as taxing authorities, copyright holders and so on. Lessig (2001, p. 247) appealed, at the time, to preserve the open code status and nature of the Internet.

The open code nature of the Internet has had some additional expressions, for instance in the evolution of some informal email correspondence codes, using signs for smiles, agreement, etc. By the very nature of the Internet as a mainly verbal communications system, literacy is much more required for its use than it is for driving, based mainly on road signs. Another informal requirement for Internet use is the knowledge of some basic computer operation. Also, some knowledge of English is almost imperative, as illiteracy of the English language implies no access to information contained in over one-half of the websites (see Hargittai 1999; W³Techs 2013). Culture and religion are some additional informal dimensions that may, in some cases, influence the extent of use of the Internet, and its open code nature, notably when religious authorities attempt to restrict access to the system or when they enforce censorship on its use. Furthermore, the use of the Internet is not only facilitated by its accessibility and affordability, but also by the capabilities of its users, as well as by their very choices of preferred uses (Kline 2013).

The Internet was considered to constitute "a metaphor for the social life as fluid" (Urry 2000, p. 40). Thus, the term *Internetness* (Kellerman 2006a) was proposed as referring to values, practices, norms and patterns within the three spheres of individuals, society and space, regarding the extensive adoption and use of the Internet. By the same token, if not used for incoming telephone calls through Voice over Internet Protocol (VoIP), the Internet cannot be considered a time-intruder for its users. In other words, the Internet facilitates its operation by users at any time of their individual choices, but it does not amount to an intrusion or intervention into the time of communicating parties, as compared to incoming telephone calls.

## Mobile communications technologies

Mobile telephone technology was originally introduced in 1906 by Lee de Forest who claimed then that, "it will be possible for businessmen, even while automobiling, to be kept in constant touch" (Agar 2003, p. 167). As noted already, the first limited mobile services were introduced in the UK in 1940 and in the US in 1947, followed by commercial introduction in 1979. Mobile phones were rapidly adopted as of the 1990s (Lacohée *et al.* 2003; Rogers 1995: 244–6).

From the perspective of individual users, the most dramatic trend in ICT innovations in recent years has been the Internet becoming fully mobile and universally available, thus permitting access to both email and the Web at any time and place, mainly through two leading technologies (Wi-Fi and 3G cellular modems). As mentioned already, this universal and instant availability of the Internet has implied a practical integration of cyberspace with physical space from the perspective of individual users (Kellerman 2010). However, as Mossberger *et al.* (2013) have strongly argued, the use of mobile broadband cannot become a substitute for fixed broadband. This is because mobile broadband, notably when installed in smartphones and tablets, is used mainly for routine communications, as well as for some basic information retrieval, whereas fixed broadband permits and facilitates much wider uses of the Internet, notably for productive and innovative purposes and opportunities. Thus, its availability should be guaranteed universally and to all social sectors, notably within urban contexts, and this should be achieved also through governmental policies, aiming at the provision of "digital citizenship" to all societal segments.

We may identify four phases in the emergence of mobile information technologies from their users' perspectives, beginning with the introduction of basic communications devices, the laptop and the mobile phone, moving through innovations of information and transmission systems, and ending up more recently with the introduction of advanced communications devices based on the previously innovated transmission systems (Table 1.2). First was the invention of the laptop PC, which was introduced commercially back in 1975, operating at the time without wireless communications. Its current penetration rate has lost some of its significance since smartphones and tablets can do much of the work of laptops. However, when comparing the contemporary popularity of laptops to that of desktops, it would be interesting to note that as of 2008 more laptops were sold in the US than desktops.

As we noted already, the mobile phone was a much older innovation than the laptop. Recently, the mobile phone has turned into the globally most widely diffused communications device, with a global penetration rate of 85.7 percent in 2011, coupled with some 90 percent of the world populated areas covered by a mobile phone signal in already in 2009. These high penetration percentages represent an even wider availability of mobile phones, given that in developing countries people may rent their mobile phones for single calls or SMSs to others as a commercial service. As of 2002 the percentage population worldwide owning a mobile line has been higher than that having a fixed line (ITU 2010, 2011).

*Table 1.2* Mass introduction and penetration of innovations for virtual mobility

| Innovation | First year of mass production | Global penetration rate in % (end of 2011) |
|---|---|---|
| *I. Basic devices* | | |
| Laptop PC | 1975 | N/A |
| Mobile phone | 1983 | 85.7 |
| *II. Information systems* | | |
| Internet | 1994 | 32.5 |
| SMS | 2000 | N/A |
| *III. Transmission systems* | | |
| Wi-Fi | 1991 | N/A |
| 3G | 2001 | 15.7[a] |
| *IV. Advanced devices* | | |
| Smartphone | 1993 | N/A |
| Netbook | 2007 | N/A |
| Tablet | 2010 | N/A |

Source: Kellerman (2012).

Note
a  Mobile broadband subscription.

Penetration levels of mobile phones were found to be related to personal income, notably in Oceania and Asia (Comer and Wikle 2008), whereas the pace of penetration has been higher in small and/or densely populated countries, permitting easier setting up of wireless infrastructure (Castells *et al.* 2007). Mobile telephony has shown evidence for a positive impact on economic growth, notably in developing countries (Kauffman and Techatassanasoontorn 2009), with special significance in Africa, since this technology is frequently the only available mobility technology there, representing a *leapfrogging process*, in which a new technology is adopted while skipping the adoption of older ones, in this case mainly the telegraph and the fixed-line telephone (Comer and Wikle 2008). Castells *et al.* (2007) noted, in their internationally comparative study, that adolescents and young adults have led in the utilization of mobile phones for SMS communications, since it has been easier for them to adopt this technology and since its use has been cheaper than that of voice calls. They further noted that with growing rates of adoption of mobile phones within given national populations, gender differences in adoption rates decline, with women tending to make more social uses of their mobile phones than men.

Two information systems have become accessible for the two mobile communications machines, the laptop and the mobile phone: the Internet and SMS. The first, the Internet, as mentioned already, became commercial and fully introduced to the general public only in 1994, and originally it was meant for use with desktops, and later also with portable computers connected through fixed wired servers. Only once wireless transmission systems have become available the Internet turned mobile, jointly and simultaneously for its email and Web components. Thus, about one-third of the global population made use of the Internet by the end of 2011, and about one-half of these users enjoyed continuous mobile broadband Internet connection via 3G (ITU 2012, see Appendix for specific data on Internet penetration for selected countries 2000–2012). Most of the Internet devices require relatively expensive terminals (PCs, laptops or smartphones), with newer tablets priced much lower. It further requires literacy for most of its uses, with the exception of visual and audio information, such as video clips and phone calls. The SMS, on the other hand, introduced in 2000, was originally invented for text and later also for video messages transmitted through mobile phones, and this is still its major application, though SMSs may currently be sent also through computers as well as through some fixed line telephones.

Though originally introduced in the early 1990s, Wi-Fi was widely adopted only as of the second half of the 1990s, since its adoption required also the development of tiny communications components installed first in laptops and later on also in smartphones, side-by-side with the need to install numerous fixed Wi-Fi antennas, each covering a limited spatial range of reception. It is difficult to assess the rate of diffusion for Wi-Fi since Wi-Fi routers are used both within and outside homes, so that its gross diffusion rate may be misleading. For late April 2010 there were reported some 295,589 free and pay-for Wi-Fi hotspots worldwide, almost a quarter of which were located in the US (JiWire 2010). The minimal transmission speed considered as broadband transmission, or high-speed transmission of data, whether through cables or through Wi-Fi, has increased over the years from 64K bps (kilobits per second) to the current widely used FCC rate of 768K bps (ITU 2003; FCC 2009).

The second mobile transmission system, 3G, permitting broadband transmission, has made it possible to use the Internet through mobile phones without the need for available limited-range antennas, but for additional charge by mobile phone companies. This technology, first introduced in the early 2000s, was available for Internet use by the end of 2011 to some 15.7 percent of the world population through mobile broadband subscriptions (ITU 2012).

Interestingly enough for the study of the growing mobile access to the Internet, the diffusion rate of mobile broadband, 15.7 percent by the end of 2011, was almost double that of fixed broadband at that time stood at only 8.5 percent of the world population. Though mobile broadband services are as of yet available to just a small portion of the global population, their percentage subscription grows fast.

The fast growth rate of mobile broadband subscription may soon turn it into a dominant technology, but its adoption rate may reach a limit in the developing world, even if we assume potential future drops in the prices of smartphones and broadband subscription, since most of the uses of the Internet require literacy. Castells (2009: 65) stated that mobile broadband is for the Internet what the electric grid has been for the provision of electric power, i.e., it permits universal distribution, but electricity use does not involve a capacity such as literacy. Currently there are still wide digital gaps in the adoption levels of mobile broadband connection: whereas in developed countries the adoption rate reached 51.3 percent by the end of 2011, it reached only 8.0 percent in developing countries (ITU 2012). Digital divides may be substantial also within countries, whether among regions or among population sectors, such as gender or socioeconomic sectors (see Castells *et al.* 2007; Graham and Marvin 1996; Malecki and Moriset 2008; Gilbert and Masucci 2011).

The availability of mobile wireless transmission technologies has been coupled, at the recent phase of mobile communications, the introduction of matching advanced mobile communications devices. The smartphone, originally introduced in 1993 and diffusing widely as of the 2000s, has actually developed into an advanced small-size laptop, including some additional features which used to be available beforehand only through dedicated devices, such as GPS for road navigation, and high-quality cameras. The introduction of smartphones was followed by the introduction of netbook and tablet laptops presenting an intermediate level of communications devices, being sized between the tiny smartphones and the relatively bulky laptops.

Several of the mobile broadband applications have been specifically developed for users on the move. One such development is extended GPS (Global Positioning Satellites) services, permitting navigation on a global scale. Another such application is a new generation of LBS (Location Based Services), providing commercial local information (e.g., on restaurants), which is transmitted to people's mobile phones, and changing along the movements of users' locations. GPS components in smartphones permit not only access to locational information but also the other way around: the exposure of users' locations to LBS providers makes it potentially possible for providers to interfere with the privacy of users, even though they are located in public spaces (Gordon and de Souza e Silva 2011).

The availability of mobile broadband jointly with the availability of proper communications devices for connection to the Internet implies more extensive and more instant uses of Internet applications, which have been developed originally for fixed broadband communications. Striking such applications are entertainment services, permitting radio and cellvision reception, and the view of streaming pictures in form of video clips and full movies. Mobile connections permit also e-work (synonymously called telecommuting or telework), e-banking, e-government, e-health, e-learning (synonymously called distance learning), and B2C (business to customers) e-commerce (synonymously termed online shopping) (see Chapter 7). In addition, SMS has turned from a mere interpersonal medium into a business-to-customers medium, as well. The British KAPOW! Survey (2005) was able to rank the top ten business types using SMS: recruitment agencies; entertainment information services; clubs and bars; Internet service providers (ISPs) and hosting companies; couriers; schools, colleges and universities; hair salons, dentists and surgeries; mechanics and body shops; charities; and insurance companies.

All these applications imply not just a widening consumption or use of the Internet by individuals, but also the widening and deepening of the production side of the information society, consisting of the production and maintenance of additional or enhanced websites, as well as the development and marketing of mobile phone applications, sold through virtual stores of mobile phone manufacturers. The availability of these applications fosters the consumption of more products and services by individual users of the Internet. Thus, a vicious cycle of growing use (or demand) and growing supply of mobile phone Internet services has emerged.

The use of mobile broadband applications may widen the multitasking pattern that has typified the use of mobile phones while on the road (see Schwanen *et al.* 2008), as well as a more flexible coordination of face-to-face meetings (Larsen *et al.* 2008). On the other hand, however, the use of some broadband applications, at home and elsewhere, has tended to develop relatively slowly at the time, since some of these applications required changes in the supply side of the system, such as legal changes for online shopping, whereas others call for adjustments of the demand for them, such as required changes in learning habits typifying e-learning (see Chapter 7).

The growing popularity of Web 2.0 applications for social networking has implied a potential permanent presence of users for both consumption and production of Internet materials. Mobile broadband has, thus, made the Internet fully turn into a routine component of daily life, without any barriers of access and without any conceptual distance between it and real space. There is no need anymore for users to move to a different arena at a special location (i.e., moving to a location of a desktop PC) in order to access and use the Internet.

The instant access to broadband services has some additional implications concerning contemporary society. First, growing virtual mobility may increase rather than decrease physical mobility, since one does not have to be tied to a desktop anymore in order to instantly initiate, receive and respond to emails and other communications, including long international telephone calls permitted now via free or low cost VoIP services. Second, the speeding up of daily activities, whether for production or for social communications, may reach now a higher level, since all communications and information media have become fully mobile, thus prompting continuous attention by users. Much before the introduction of the commercial Internet, Virilio (1983: 45) called our era *"the age of the accelerator,"* and this nature of contemporary society has been accentuated time and again, notably regarding car driving, or accelerated and personal physical mobility. "Speed is the premier cultural icon of modern societies.... Speed symbolizes manliness, progress, and dynamism" (Freund and Martin 1993: 89, see also Kellerman 2006a). Third, the blurring of separation between work/business and leisure which has typified the spheres of work and home in recent years may continue to intensify, since both work and social activities can be easily performed when away from both office and home. Thus, from the perspective of the location of work activities, Castells (2001: 234) referred to *nomadic workers*. The blurring of boundaries between leisure and work activities is further amplified, since for both activities the same equipment, software and channels are used (see Kellerman 2006a). At the sphere of social relationships, Licoppe

(2004) recognized an emerging pattern of continuous "connected relationships" through various media of electronic communications, so that "the boundaries between absence and presence eventually get blurred" (p. 136).

The use of either mobile phones or wireless Internet connection has an implication also for the public sphere, in that it may blur the distinction between the private and the public, as well as that between indoors and outdoors (Kopomaa 2000). Whereas telephones and computers were traditionally considered devices to be used indoors and involving some privacy of communications, wirelessness has implied less privacy and a change of social boundaries, given the acceptance of communications activity in the public sphere (see Kellerman 2006a). Thus, the contemporary social environment is characterized by a blurring between the public and the private through the use of mobile phones, into what Sheller (2004) termed "mobile publics." Furthermore, mobile surfing of the Internet accentuates placelessness, or less attachment to places of residence, which may arise through web surfing at large. This may develop since virtual mobility via the Web implies exposure to remote places while being physically located in a fixed place or while being on the road. "The contradictory experience of being somewhere and nowhere at the same time is perhaps the most obvious cognitive dissonance resulting from the use of the WWW" (Kwan 2001: 26, see also Kellerman 2002: 39–41, 49).

## Auditory geography of the Internet

Our discussion so far has related to the Internet, at least implicitly, as a visual space. However, the Internet has to be assessed also from an audio perspective, as presenting its own auditory geography. Contemporary culture at large has been assumed to be more visual than auditory (Ong 1971), though, as we will see, sound too possesses numerous geographical dimensions. Rodaway (1994), proposed the term *auditory geography* as referring "to the sensuous experience of sounds in the environment and the acoustic properties of that environment through the employment of the auditory perceptual system" (p. 84). More recently, Kanngieser (2012: 337), following Nancy (2007) and Matless (2005), preferred the notion of *geography of the voice*, rather than auditory geography, referring to "the ways that voices are shaped by, and shape, worlds and spaces." Most of geographers' attention in this regard has been paid to music, as recently reviewed by Kanngieser (2012). In the following paragraphs we would like to propose some elements for a sonorous/sonic geography, with special attention to the Internet, noting its basic elements of sound, silence and noise and extending it to the virtual space of the Internet.

Sound is no less basic to human sense and experience as compared to sight, consisting of a sonic spectrum ranging from loudness, or noise, to silence (LaBelle 2010: 79). Silence, as such, may be viewed and experienced either as a sound by itself or as an unheard sound element equivalent to the invisible regarding sight. Interestingly enough, recent writings on the study of sound frame their notions within concepts borrowed from geography, commenting also on similarities and differences between sound and sight. Thus, *soundscape* was proposed as an equivalent notion to landscape: "The 'soundscape' is the sonic environment which surrounds the sentient" (Rodaway 1994: 86; see also Smith 1994), and "we may speak of a musical composition as a soundscape, or a radio programme as a soundscape or an acoustic environment as a soundscape.... A soundscape consists of events heard not objects seen" (Schafer 1994: 4). A place is generated by the temporality of the auditory (LaBelle 2010: xvii), constructing ever-changing sonic maps (Voegelin 2010: 136–7). *Sonography* was considered as equivalent to geography dealing, for instance, with the decomposition of a soundscape similar to geography's decomposition of a landscape: "the sound-scape corresponds to the whole structure of the text, while the sound object corresponds to the first level of composition: words and syntagmas" (Augoyard and Torgue 2005: 7). The combination of noise and silence implies that "an auditory geography exists [then] within the very meeting or interweaving of noise and silence, forming a continual articulation of what is permissible" (LaBelle 2010: 47).

In everyday life, the highly dynamic and constantly changing soundscapes, and the somehow less dynamic landscapes, are interwoven into each other. "The environment can be considered as a reservoir of sound possibilities, an *instrumentarium* used to give substance and shape to human relations and the everyday management of urban space" (Augoyard and Torgue 2005: 8). Furthermore, spatial configuration and the aural sphere may provoke, jointly and separately, an *anamnesis* effect of personal remembrance (Augoyard and Torgue 2005: 24). However, the practical relations between built space and sound are more complex. On the one hand, sound can move across invisible and sometimes even through material boundaries (LaBelle 2010: xxi), thus having the "capacity to disintegrate and reconfigure space" (Connor 1997: 206), and more generally

> the singular space of the visual is transformed by the experience of sound to a plural space; one can hear many sounds simultaneously, where it is impossible to see different visual objects at the same time without disposing them in a unified field of vision.
>
> (Connor 1997: 207; see also Ingham *et al.* 1999)

Still, however, "constructed spaces create divisions in the sound environment influencing the propagation of sound sources by orienting movement and progression through theses spaces" (Augoyard and Torgue 2005: 30).

Mobility technologies are not only significant components within the material and visible contemporary urban environments but they present specific sound elements as well. Human attempts to compress time and space using physical mobility technologies have brought about the most extensive sources of noise in the city, caused by terrestrial and aerial traffic (Keizer 2010: 15). This noise amounts to a drone in city life at large (Augoyard and Torgue 2005: 40), but still each city has its own drown tone (Prochnik 2010: 110). Communications technologies too have presented soundscape/landscape aspects. Thus, the telephone and the radio released the presentation of sound from its point of origin, while the phonograph and following sound recording technologies released sound from its time of origin (Schafer 1977). The virtual space of the Internet, from this perspective, presents sounds which are released from both their times and places of origin. The Web constitutes, therefore, a virtual soundscape, fully integrated and synchronized with its being a virtual landscape, and dynamically changing by its users!

Sound may be used to dominate and control space (Rodaway 1994: 113), but this is irrelevant for mostly silent cyberspace. The conditions of silence in the use of the Internet may be assessed from several perspectives, such as sound continuum and performance, which we will briefly review now. First, the Internet can be assessed as operating along the continuum silence → noise, namely as a silent medium, which can be turned into a noisy one through speakers by clicking just once or twice on a virtual screen button. The Internet constitutes foremost a visual experience, through its rather textual email component, as well as through its visually more diversified Web component, which includes textual, graphic, still and streaming video modes of information, all of which constitute visual information entities. The Internet provides users with the choice of using sound or refraining from it, by the type of use and by personal preferences. In VoIP voice communications as well as in telephone calls and in listening to music and movie sounds, sound is either the sole or an integral part of the application. The Internet further permits users to *consume* sound individually by listening to it through earphones. The Internet obviously permits users to *produce* sound or noise by talking to a communicating party. The Internet may, thus, be viewed as a combination of silence and sound, which can be fine-tuned individually by users depending on the type of use and the time of operation.

Second, the Internet can be assessed from the perspective of sound performance. Several contemporary daily or weekly consumption activities have now two alternatives: a noisy public one in real space and a private and quiet one via the Internet, such as socializing and shopping. For the public and rather noisy option in real space it was noted that, "noise is used extremely as a marketing tool (bars, restaurants, public spaces in general, radio, television, film), as a way of stimulating consumption" (Sim 2007: 30). This culture of noise has its effect also on the Internet. Thus, even in the silent (i.e., non verbal) options of Internet use, noise may burst into silent websites through audio-visual advertisements. Furthermore, Internet performance, including both email correspondence and Web browsing, may increasingly take place within noisy real space contexts, such as parks, cafés, airports and so on, notably since the distribution of public and commercial Wi-Fi networks has expanded. Thus, whereas Internet activities might be silent by their very nature, they do not always require silent physical surroundings for their very operation by users. This is not to say, though, that all Internet activities can be performed within a noisy context. However, the preparations for activities that require silence can take place within a noisy context, such as the downloading of scientific literature, for example, in preparation of its reading and study.

In their choice between silent and noisy modes of Internet use, users adhere to Sim's call for silence to be a choice rather than an imposition (2007: 38). This same rule applies also to the very choice people can make between the use of the Internet as compared to a possible use of other, noisier, communications media (e.g., mobile phone calls) or shopping activities (e.g., mall shopping). The contemporary personal autonomy of individuals permits persons to decide whether to manage their lives, at any time and for any activity, through either silence or noise, or alternatively, when and where to be part of a noisy environment (e.g., urban streets, malls, discos, etc.), or to prefer a rather silent one (Internet). This demand side of noise and silence in the use of the Internet is coupled with a supply side: contemporary society has developed and it offers technologies for silent uses side-by-side with those for noisy ones.

An important example of social choice between a corporeal noisy action and a virtual silent one is personal mobility (Kellerman 2006, 2012a). Corporeal mobility always involves noise, such as through driving, as well as in riding public transportation, or when flying. Personal virtual mobility, on the other hand, may be noisy when speaking in calls made through communications media, but it may also be silent when writing emails and SMSs. Such physical silence in the written interaction through the Internet and through mobile phones is achieved instantly, as compared to the silent but rather lagging interaction through postal letters. Once again, silent virtual mobility may replace some noisy corporeal one, but in most cases silent virtual mobility is rather complementary to physical noisy one, and it takes place either before and/or after face-to-face meetings involving some physical and noisy mobility.

## "Action space" and the Internet

It seems that the social sciences have not explicitly provided any general notions for human action vis-à-vis action potential and action capacity. The specialty of "action research" is rather devoted to the generation of knowledge "for the express purpose of taking action to promote social change and social analysis" (Greenwood and Levin 1998: 6). In geography, though, the study of routine activities of individuals has been based on a distinction between two spatial spheres. The first and wider one is *action space*, and it was defined at the time as "an individual's total interaction with and response to, his or her environment" (Golledge and Stimpson 1997: 277). Within this wide geographical sphere of *potential* action of individuals lies the more restricted *activity space*, "defined as the subset of all locations within which an individual has direct contact as a result of his or her day-to-day activities" (Golledge and Stimpson 1997: 279), constituting the *actual* spatial range of people's activities. Dijst (2004) referred to action space as the *potential action space*, to activity space as the *actual action space*, and added to these the *perceived action space*, relating to the areas known by individuals as action spaces.

These clear-cut differentiations between potential and actual daily spatial spheres of action and activity respectively have been somehow blurred when individuals have been exposed to the new possibility to perform numerous activities virtually and, thus, globally through the Internet. Frequently Internet users are not even aware of the specific locational anchoring in real space of the activities that they perform through websites being available globally. Hence, the contemporary action space of individuals, referring to the potential geographical range of their activities, has become global, and hence our attempt to assess the Internet as a second action space for individuals, unrestricted by the traditional constraints of location and distance.

Side-by-side with their potential global action space, individuals' practically used activity space has extensively widened, as well, even if its extent might frequently be still mainly domestic rather than global. The activity spaces of individuals have widened through the use of the Internet as compared to the local range of activity before the emergence of the Internet as an action space. The traditional activity space has been restrained by physical access to facilities or by reasonably priced telephone access to mainly local facilities and locations, whereas the Internet offers free access to anywhere. However, even under these more flexible conditions, individuals still assess their potential mobility for each activity they need or would like to perform in both real and virtual spaces (Kellerman 2012a, 2012b). The increasing geographical spread of potential destinations for human actions was termed by Giddens (1990) as *distanciation*, and this applies to a variety of human activities, such as person-to-person communications, the retrieval of information, and the ability to work and to buy services globally.

Cyberspace is not only an action space without distance and location barriers, nor is it merely a new world of information. It has further been viewed as constituting also a dense concept of converging technological environments, human minds, personal motivations and a generation source for all kinds of human artifacts (Dodge and Kitchin 2001). We will attempt to highlight these dimensions in the next chapters. Furthermore, in his study of mental total institutions, Goffman (1961) argued that physical structure and architectural settings have an important impact on actions and interactions of inmates. Davis (2010) has taken this argument one step further by claiming that this may apply also for virtual spaces, so that the specific design and organization of a website may have an important impact on patterns of human action and interaction within it.

The focus of this volume on virtual action space is mainly on the diversion of activities from real space to the virtual one, as far as daily deeds are concerned. As such, our focus in Parts II–III will be mainly on consumption activities by individuals, reflecting a variety of human needs and motivations. Ahas *et al.* (2010) viewed people's homes and working places as major "anchor points" in their activity spaces, whereas all the others, such as shopping and social as secondary ones. We have not adopted this differentiation, since our objective here is to present and assess the diversion of people's activities at large for real space to virtual one. Yet another classification, proposed originally for trip purposes, is relevant for the classification of action spaces as well: mandatory or subsistence (work and study, see Chapter 7); maintenance (e.g., shopping, banking, health, government services, etc., see Chapter 7); and leisure (e.g., touristic travel, social networking and gambling, see Chapters 7–9) (Meyer 1999, see also Järv *et al.* 2012).

As we will note in Chapter 7, the potential moving of the arena of work or production, solely or mainly, into the virtual sphere has not been massively adopted. The Internet has permitted, though, the emergence of a new and third mode of action, in addition to production and consumption, a mode that is entirely based on Internet communications and virtual action space. Bruns (2008) termed this new mode as *produsage*, referring to the joint, open, fluid, open-ended, uncoordinated or unmanaged production and consumption of knowledge and information through dedicated websites and forums, such as Wikipedia which permits readers to add to its entries and blogs and chat platforms which are based on continuous consumption and production of information among networked individuals. We will explore online social networking in Chapter 8.

## Conclusion

In this chapter we reviewed and highlighted cyberspace and the Internet from numerous perspectives. First, we defined cyberspace, focusing on its nature, then elaborated on its two classes of information and communications spaces, and on its cognition as a virtual entity. We then moved to a comparison between real and virtual spaces in their basic dimensions of organization, movement and users, followed by an exploration of the practical relations between the two spaces. These three rather basic discussions on cyberspace presenting definitions and categorization, and conceptual and practical comparisons with real space, paved the road for a brief presentation of the Internet, its foundations and its characteristics as an open and universally accessible information system. The Internet has become increasingly accessible through mobile information technologies, which we described and assessed in terms of the more extensive actions that they permit through mobile communications devices. We ended our introductory discussions on cyberspace and the Internet by adding to the normal discussions on the visual dimension of the Internet some comments on the auditory geography of the Internet as compared to that of real space. Following this introduction of the foundations of the Internet and its accessing we paid some attention to the notion of action space and presented the classical definitions for action and activity spaces, assessing them vis-à-vis the Internet.

A repeated notion in this chapter was the relationship between real and virtual space, and we noted several such relations. We discussed the practical integration and convergence of cyberspace with real space; the two spaces being interlinked, alongside noting that cyberspace is embedded in real space and is organized like real space, as well as representing real space. This variety of associations points to rather complex relations between the two spaces which we will further highlight in the next chapter, devoted to theoretical perspectives of the Internet as a second action space. It may further point to a still emerging and as of yet unsettled development of the Internet as an action space, interrelated with the more traditional real space, a perspective which will accompany us throughout the following chapters.

Three additional possible relations between real and virtual spaces were recently proposed by *The Economist* (2012): that the digital world reshapes the physical one; that the real and the virtual worlds are separate from each other; and that the physical world shapes the virtual one. One may argue that all three possibilities are true and each being relevant to different spheres of the virtual world. We will explore these relations in later chapters.

# 2 Theoretical perspectives on the Internet as second action space

The Web is a special kind of space.

(Weinberger 2002: 33)

In the previous chapter we have become acquainted with the basic features of virtual space per se, as well as with their comparison to those of real space. In this chapter we will delve a bit deeper into the nature of virtual space as compared to that of real one, in an attempt to explore some basic meanings, significances and workings of these two spaces, notably from the perspectives of their constitution as social and action spaces. Hence, four perspectives will be offered in this chapter: attributes of real and virtual social spaces; experiencing real and virtual social spaces, notably under the assumption that virtual space has turned into action space; social forces in the emergence of virtual action space; and time geography aspects of virtual action space.

We will try to show that the numerous attributes, interpretations and metaphors which have been suggested for real social space are mostly relevant for virtual space as well, though sometimes in different ways than for real space, and this despite the lack of geographical scales in it, such as city and region. The issue of the experiencing of virtual action space will be discussed vis-à-vis Lefebvre (1991; developed further by Harvey 1989 and Soja 1996), whose notions of space classes seem to be challenged by the emergence of the Internet as an action space. It will be argued that virtual action space is a novel entity somehow differing from real space in its experience, being more flexible in both its very construction and in people's moving through it.

The social meanings of the construction process of virtual action space and its following use will be highlighted through structuration theory (Giddens 1990), which accentuates the vicious cycle between human agency and social structures. It will be argued that technology serves as a mediator beween human needs and actions as reflected in real space, on the one hand, and business motives, on the other, shaping gradually a virtual action space. The distanciation of human action space has reached its utmost in virtual space, side-by-side with time-space compression, which has increased to unprecedented levels in its use. Finally, virtual action space will be assessed by the time-space prism and the action constraints that Hägerstrand (1970, 1973, 1975) proposed at the time for real action space. It will be argued that the time-space prism does not exist with universally available mobile broadband use of the Internet, and that several of the time-space action constraints proposed for real space have either changed or have been nullified for virtual action space.

## Social space attributes

*Space* in general may be viewed as one of the primal notions of geography (see, e.g., Kellerman 1989), and as such, its constitution, meaning and expression being enormously varied (see, e.g., Simonsen 1996). Space as a dimension of wide geographical extent does frequently carry either physical or abstract connotations (see Gregory *et al.* 1994: 8), but it is further loaded with a variety of social meanings as well, and these may apply to both real and virtual spaces. We will focus our attention to social space in this regard.

Numerous attributes, interpretations and metaphors have been offered over the years for material social space: distance (e.g., Ilchman 1970), barrier (e.g., Harvey 1989), dimension (e.g., Ullman 1974; Massey 1992), resource (e.g., Kellerman 1989), social value (e.g., Sack 1980), material entity and product (e.g., Lefebvre 1991), landscape (e.g., Cosgrove 1984), place (e.g., Giddens 1990: 18; Entrikin 1991; Merrifield 1993), container (e.g., Kellerman 1989), production force (e.g., Swyngedouw 1992), text (e.g., Barthes 1972; Duncan and Duncan 1988), symbol (e.g., Lefebvre 1991), process (e.g., Merrifield 1993) and organizational framework (e.g., Soja 1989).

These attributes, interpretations and metaphors may all be considered as relevant for virtual space as well. Thus, virtual space may constitute: a resource and a production force for e-commerce; a text and a symbol in its very nature; and as an organizational framework in almost all its uses and applications. As we noted already, some would argue for it being also a landscape, a place and even a social value (see Dodge and Kitchin 2001 for detailed discussions). Virtual space can be looked upon by far as an imagined space of representation through its virtual imitation or virtual description of real spaces and places.

Yet another attribute of social space, and a quite ambiguous and complex one in the emerging individual uses of virtual space, is distance, which received much publicity at the time through the provocatively entitled book *The Death of Distance* (Cairncross 1997). Several additional writers have noted the decline in the importance of distance in regard of telecommunications technologies at large (e.g., Gillespie and Williams 1998; Gillespie and Robins 1989; Atkinson 1998; Brunn and Leinbach 1991; Castells 1989; Negroponte 1995). However, this perspective relates to measured distance in real space as losing its crucial importance through virtual electronic communications.

Yet another view of distance would be to consider its possible role within virtual space despite the lack of and irrelevance of measured distance in it. Weinberger (2002: 45) noted in this respect that "distance on the Web is measured by links," and "links are all that holds the Web together; without links, there is no Web" (p. 54). Thus, distance over the Internet may be measured by the number of clicks needed by users to reach any specific page or website. Distance is, thus, an indicator of time, convenience and level of complexity.

A third interpretation for distance in the use of the Internet would be the notion of access to the Internet, or the distance users have to pass on their way *to* the Internet rather than distance *within* it, an issue which we discussed already in the previous chapter. This distance may be expressed and measured by infrastructure features and limits, such as bandwidth and the extent of availability of mobile broadband. This latter notion of distance as users' access to the Internet may reflect digital gaps among individuals and social sectors and regions in the ability to access the Internet, thus bringing about a rather restricted, and thus inefficient use of the Internet by some users, and a complete avoidance of its use by others. We will further refer to this notion in a later section focusing on time-geography.

The numerous attributes, interpretations and metaphors attributed to real social space have been expressed, mediated or embodied along several geographical and social scales, extending from the local-urban, through the regional and the national to the global (see Massey 1992). These spatial scales do not merely constitute a convenient professional, administrative or political classification of spatial size, but they present also routine human spatial practices and experiences, perceived and organized along these scales (see, e.g., Shamai and Kellerman 1985). In virtual space this geographical differentiation by scale seems irrelevant, since access to websites and their use may almost extend globally, without a differentiation between websites, which are supposedly accessible only within specific cities, regions or countries. Furthermore, the Web as a system, is not organized by any geographical or by any other scales. Still, the geographical dimension of scale for virtual space is there in somehow complex and mixed ways. Thus, while the Internet spans globally, it may, in many cases, visually present local and regional images and information, as well. Furthermore, websites may be written in domestic languages, so that access to the websites is also mostly or only domestic. The Internet thus constitutes a "different human experience of dwelling in the world; new articulations of near and far, present and absent, body and technology, self and environment" (Crang *et al.* 1999: 1).

Multiple junctions may emerge between attributes of real and virtual social spaces related to or mediated through human relations. For example, urban space may be considered as a text or as an imagined space for the design of a website in virtual space, but urban space as land may further be considered a resource for the real location of an information business of a webmaster who designs websites. Contemporary social space constitutes, therefore, a complex double arena of real and virtual spaces, which may carry several meanings simultaneously at specific times, places and spatial scales, changing by different groups of people and individuals.

## Social space experiencing

Let us first examine some social views on the experiencing of real space and its construction, and then we will attempt to examine these views as far as virtual space is concerned. A most insightful socio-spatial classification focusing on the human experiencing of social space was proposed at the time by Lefebvre (1991; see also Merrifield 1993 and Kirsch 1995), differentiating among material spatial practices, the representations of space, and the spaces of representations, jointly constituting human social spatiality. Harvey (1989: 220–1) interpreted these relations as relating to human (direct) spatial experiences, and the individual perception, and imagination of space, respectively. Soja (1996) interpreted Lefebvre's three classes a bit differently, by claiming that they rather constitute a trialectic of human perception, conception and living of space, respectively. Material spatial practices, or *Firstspace*, á la Soja (1996: 6, 66), are those practices which are performed by all individuals in real space and which further imply spatial perception by them. These geographies of our lifeworlds constitute, among other dimensions, "the complex spatial organization of social practices that shape our 'action space' in households, buildings, neighborhoods, villages, cities, regions, nations, states, the world economy, and global geopolitics" (Soja 1996: 75). To these geographical arenas of material social practices in real space one may add these days the Internet, as well, as we will argue in a following paragraph.

The representation of space, was coined by Soja as *secondspace*, "conceived in ideas about space, in thoughtful re-presentations of human spatiality in mental or cognitive forms," normally by professionals of space, such as planners and urbanists (Soja 1996: 10, 66). *Thirdspace*, namely the spaces of representation or lived space, was referred by Soja to a re-combination and extension of the first two spaces, but simultaneously constituting a class by itself, "a space that stretches across the images and symbols that accompany it" (Soja 1996: 6, 67). This thirdspace was attributed to artists, writers and philosophers who use space in a "warm" manner as compared to the "cold" manners of the use of space made by professionals employing presentations of space as secondspace (Soja 1996: 30). Thirdspace as a moment of social space constitutes further "the space of radical openness, the space of social struggle" (Soja 1996: 68).

Castells (2000: 441) argued for the oneness of society and space in his statement arguing that "space is not a reflection of society, it is its expression. In other words: space is not a photocopy of society, it is society." Moreover, one may argue also the other way around: society is not a reflection of space as it constitutes its expression, so that human practices create social spaces which on their part are conceived and perceived by the spatially practicing individuals side-by-side with fellow individuals. Such a view downgrades a bit the sharp division of space into classes per its experiencing from a social perspective. Soja (1996: 77–8) preferred, though, a one-way relationship regarding the causal relationships between society and space, noting that society has been assumed as bringing about the production of material space but ignoring the possibility for an opposite causality. The argument for a two-way oneness of society and space may apply to virtual space as well. We noted already in the previous chapter and we will further note in the following paragraphs that the relationships between production and use for virtual space are similar to those for real one.

Back in 2002, when the Web began to emerge as a second human action space, virtual space was presented as constituting a space of representation, reflecting real space (Kellerman 2002: 37–8). Figure 2.1 presents the three dimensions of social space and the relations among them, assuming the human use of both real and virtual space and assuming that virtual space is a representation of real space. Thus, material spatial practices constitute flows, transfers and interactions taking place in real space, whereas virtual space is presented as a space of representation/imagination. These two spaces were viewed as being mediated through codes (information) and knowledge, or the representations of space. The construction of virtual space was, thus, assumed to be embedded within real space through a double experience: an *operational* one relating to the transfer of activities from real space to virtual one, coupled with a *metaphorical* experience, through the use of geographical language, symbols and tools in the construction and use of cyberspace. The cumulative experience in using virtual

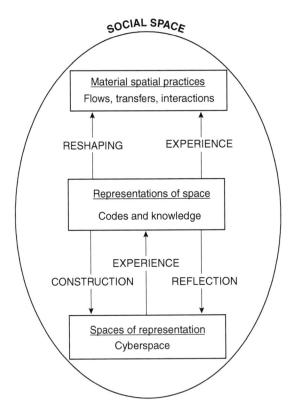

*Figure 2.1* The three dimensions of social space (source: Kellerman 2002, Figure 2.2).

space may have led, again through human knowledge and codes, to a possible reshaping of real space, in terms of locational patterns of both production and consumption (see Kellerman 2002; Wilson *et al.* 2013). Information and knowledge constitute, therefore, mediating forces between the construction and reshaping of real and virtual spaces.

The constant discourses between real (material) and virtual (imagined) spaces accentuate the oneness of the material, the perceived and the imagined, as far as social space is concerned. Harvey (1989: 219) noted such an option, before the introduction of the Internet, when he stated: "The spaces of representation, therefore, have the potential not only to affect representation of space but also to act as a material productive force with respect to spatial practices." This view resembles Castells' (2000: 441) view on the oneness of society and space, which we noted already in a previous paragraph.

The view of the virtual Web space as a social space so far assumed its being a reflection of real social space, and it is, thus, in line with the numerous relationships between the two spaces which were presented in the previous chapter. However, the more recent development of human ability to perform widely on the Web through the maturing of the Web as a second action space call for looking at it also as a social space by itself, beyond its being merely spaces of representation for real space. Let us, therefore, compare real and virtual spaces from the perspective of the three social dimensions developed originally for real space and presented previously, with the aid of notions proposed by Weinberger (2002) and by Meishar-Tal (2006). This comparison is presented in Table 2.1, and elaborated on in the following paragraphs.

The basic societal dimension for both real and virtual space is the routine spatial practices of individuals, whether of a social or economic nature. In real space, as we noted already, each of these practices is performed within a distinct

*Table 2.1* Dimensions of real and virtual social spaces

| Dimension | Real social space | Virtual social space |
| --- | --- | --- |
| *Spatial practices* Flows, transfers and interactions | Performed by individuals in distinct geographical scales, mainly locally | Performed by individuals in instantly integrated spatial scales, dominated by globalization, using geographical language |
| *Representations of space* Codes and knowledge | Designed and studied by spatial sciences (e.g., planning); dominated by distance and terrain of pre-existing space | Designed by computing specialists; dominated by networking and linking; space is created only through the construction of websites |
| *Spaces of representation* Ideas, images and symbols | Presented by artists, writers and philosophers | Designed by graphics specialists aiming at imagined materialization of virtual space |

geographical scale, but mainly within the local one, though travel to other places is obviously a frequent or less frequent option. In virtual space, through the Internet-based Web, activities are marked by their globalization (Meishar-Tal 2006). Globalization may be even part of a seeming local virtual activity, such as sending a post through Facebook to local friends only, since the transmission of this post might be activated by servers located in one or in numerous foreign countries. Needless to say that many virtual activities may knowingly be global in terms of the accessed organization or individual, and that accessing websites is the same whether they are domestic or not. As we mentioned already, the use of the virtual space of the Internet attempts to imitate the experience of action in real space through the use of geographical language, employing words such as "site," "home," "browsing," "going," etc. "Space isn't a mere metaphor. The rhetoric and semantics of the Web are those of space. More important, our *experience* of the Web is fundamentally spatial" (Weinberger 2002: 35).

Representations of real space are carried out through codes and knowledge which determine human social practices and are influenced by them, and which are designed by professionals of space, such as planners and architects. In their design of spatial facilities such as buildings and roads, planners and architects are most basically restrained by distances between facilities and their potential users, as well as by topographical terrain of pre-existing space, acting as a container for newly-designed facilities. In virtual space these two factors of distance and terrain are nonexistent and they are replaced by two other basic factors: linking and networking. Linking refers to the linking offered from one website to others, and which constitute the heart of the Web: "Distance on the Web is measured by links" (Weinberger 2002: 45); "Links are all that holds the Web together; without links, there is no Web" (Wenberger 2002: 54). By its very definition, there are no pre-existing container-like spaces within virtual space, so that spaces are created a posteriori only, through uniquely addressed websites (Weinberger 2002: 44–5). The second factor of virtual space that enables or delimits its use by individuals is networking, notably as of the introduction of Web 2.0 in the early 2000s (Meishar-Tal 2006). Whereas linking connects users with information and action arenas, networking connects among actors or individuals. The level of networking differs among users and this level of networking may determine the extent of exchange of individual information among users, as compared to institutional, economic and other impersonal information and activities permitted by websites through linking.

Spaces of representation for real space refer to artistic and literal descriptions of space as well as to revolutionary ideas regarding its use and organization. In virtual space, artistic descriptions are next to impossible, since there is no lasting experience of exposure to a specific website and obviously there is no physical bodily living and experiencing in it. On the other hand, though, the construction of websites involves not only contents or information and not only the use of clicks/directions, but it requires also the production of enveloping, comfortable and inviting, designs which obviously differ from one website to another. Such designs, produced by graphics specialists, aim foremost at the "realization" of virtuality for its users (Meishar-Tal 2006). One may recognize also another major difference between the two classes of space regarding the politics of space. Whereas in real space there might exist physical or hidden barriers preventing entrance to specific places and spaces, the Web, in principle at least, provides "doors" for entering "other" spaces, through links which do not require the permission by website "owners" or creators (Weinberger 2002: 52–3). Still, however, entering some websites requires registration and/or payment, and for others access might be prevented by governmental censorship (see Chapter 9).

## Social forces in the emergence of virtual action space

Following our discussion on the experiencing of social space in virtual space, we may now turn to a brief exploration of social forces in the emergence of virtual action space. Giddens' structuration theory (e.g., Giddens 1981, 1990) proposed a general interplay between several major forces in the shaping of societal developments and patterns. The first two forces are socioeconomic ones or the infrastructure, consisting of systems of rules and institutions, which are coupled with social superstructures, referring mainly to religion and culture. The third social force is human agency or individual action. The theory suggests that daily human agency occurs under the impact of socioeconomic structures and superstructures, while this human agency may cumulatively bring about some structural changes.

A major social crossroads has quite swiftly emerged with the introduction, back in 1994, of the Internet in its current form, consisting of Web and email as commercial communications and information products/services. The commercial nature of the Internet has expressed capitalist desires for new ways of profit generation. At the same time, however, this new medium has reflected human needs and desires in the modern and postmodern era for speedy action, something that Virilio (1983: 45) called "*the age of the accelerator.*" The Internet has become the major medium for personally and instantly available information and communications based on computers and fixed telecommunications systems. Technology may, thus, be viewed as a mediator between human needs and actions, on the one hand, and business, on the other. These three forces of profit seeking, the desire for speedy action and communication and information technology (IT) have jointly and gradually shaped a virtual action space.

Gradually too the business world created three lines of business activities vis-à-vis the relationships between real and virtual spaces: one type of activity is that which can be performed in both real and virtual spaces and are located simultaneously in both of them. A major example in this regard is banking, with traditional bank branches still located in real space, side-by-side with globally accessible websites of banks, avoiding the use of specific branches. A second type of business is products and services are those that can be offered in both spaces, but there emerged companies which have specialized in sales over the Web only, such as Amazon, which does not have any stores located in real space, but there are numerous book companies which offer both online and in-store sales, and others which offer in-store sales only. A third type of business is a company that offers products and services that can be offered only through the Internet. Examples are information services such as the search engines and satellite images offered by Google, and the networking services available through Facebook and Twitter. These virtual business types, or the introduction of commercial services for a more efficient and comfortable use of the Internet, have emerged along several facets of human agency in the use of the Internet. First has been the very growing use of emails and virtual forums for social interaction, eventually yielding Web 2.0 in the early 2000s. Second has been the growing ease of use of virtual space when its graphic and linguistic organization imitated the traditional ways of conduct in real space, and, third, the desire for ever speedier action, pushing technology towards faster and faster communications, which on its part, was coupled with technologies enabling full mobility of access (see Kellerman 2012a and Chapter 1).

These developments have meant an amplification of two trends that emerged before the introduction of the Internet: *distanciation* and *time-space compression*. Distanciation was originally defined as the "stretching" of the spread of social systems in time and space (Giddens 1990), mainly through the development of culture, such as through writing, which permitted the transmission of ideas and information over space and time, and mobility technologies that facilitated such transmissions over space. Though distanciation refers to societal trends and processes, it is of significance also to individual daily spatial actions. Contemporary virtual mobility technologies bring about the "stretching" of social systems almost to their utmost, thus permitting extended distanciation of reach to users of the Internet, turning it both potentially and practically, global.

Time-space compression was defined as the "compression of our spatial and temporal worlds" (Harvey 1989: 240), or a "pull" mechanism, induced by contemporary telecommunications technologies. For example, a chat taking place between Australia and the UK implies that one of the two communicating parties may be awake late at night or working at that time, so that both time and space differences between the two parties and locations have been compressed. Time-space compression may be viewed as both an outcome and a cause for distanciation: extended distanciation through mobility technologies calls for more intensive time-space compression and vice versa: extended time-space compressions may lead to an extended distanciation. These trends do not relate to all individuals equally in the developed world. Thus, there are those individuals who sense this compression only passively or indirectly, if they are people who are forced to move by their employers or by the search for employment opportunities, or if they are immobile people because they serve other more mobile ones. Thus, there are others who are in charge of this compression through their handling of the local/global transfers, notably those of capital and information (Massey 1994: 149).

## Time geography and virtual action space

Hägerstrand (1970, 1973, 1975; see also e.g., Kellerman 1989) developed time-geography as a tool for tracing the time-space paths of individuals and their activities through time-space prisms, which depict daily travels for distinct activities and the lengths of stay in specific locations for each activity. He further outlined several constraints for human mobility in time-space and for human action through it. It is of interest to examine Hägerstrand's original notions as potentially fitting human action in virtual space as well.

Yu and Shaw (2008) extended Hägerstrand's time-space prisms for activities in virtual space. The spatial dimension that they used for virtual action space was users' accessibility to the Internet from real space, measured by the distance of actors to a wired computer, rather than by distance from one activity to the next one as in Hägerstrand's original model. This change was, obviously, forced by the lack of distance between activities or websites in virtual space. Thus, *space-time lifelines* were proposed by them as referring to accessing virtual space from wired computers, and the spatial dimension measured by the travel needed for reaching a wired computer. *Space-time life cylinders* were defined, referring to wireless access to virtual space with the cylinder delimiting the possible access range through specific Wi-Fi antennas. However, the growing use of mobile broadband through cellular universal connections, as a preferred mode of access to the Web in many if not most cases in recent years, nullifies the very existence of time-space prisms when its spatial dimension is defined by access to virtual space.

Hägerstrand (1973) further defined three families of constraints for human action in time-space:

*Capability constraints* that related, for example, to the limited reach of the human body, a constraint that has been relieved with extended reach through the Internet, or to the need to devote time for sleeping. Dijst (2004) suggested one should view the availability of proper devices and knowledge of their operation as capability constraints for the use of IT.

*Coupling constraints* refer to the cooperation and coordination between two or more people that is required in order to perform certain tasks. This type of constraint has been extended in some way in virtual space actions. Since contemporary instant and global communications permit extended cooperation among actors, co-performers of shared activities require careful coordination of such activities.

The third group, *authority constraints*, refer to prohibitions of entrance and movements by individuals in real space. For virtual space we noted already the free linking possibilities among websites, permitting extremely wide mobilities in virtual space. Still, however, entering and using some websites may require subscription by users followed by identification, whereas the use of other websites may involve a charge (see also Dijst 2004).

In yet another place, Hägerstrand (1975) proposed eight constraints for human action in time-space:

1    Indivisibility of human beings.
2    Limited length of human life.
3    Limited ability of humans to perform more than one task at a time.
4    Duration needed for every task.
5    Time consumed for each movement between points.
6    Limited packing capacity of space.
7    Limited outer size of any terrestrial space.
8    Past rooting of every situation.

Of these eight constraints for action in real space, the first four apply to virtual space as well, since they all refer to the limitations of humans as such and more particularly as space users. However, the last four constraints, which refer to space and time, seem to be irrelevant to human actions in virtual space, since no movement time is involved there, nor are there any limited spatial capacity and outer size of spatial units. Past rooting of situations is also difficult to assume for ephemeral appearances of screened spaces. Thus, human actions in virtual space may turn out to be most flexible for both suppliers and users of virtual activities, possibly making them also more attractive for both suppliers and users (see also Schwanen and Kwan 2008).

## Building blocks for virtual social action space

We may now suggest several basic building blocks for virtual social action space, relating to both space production and space consumption, and arranged along a potential sequence of action, leading from production to consumption:

1    *Production of representations of space*: The production of websites by computing specialists, using codes and knowledge, and dominated by networking and linking principles.
2    *Design of spaces of representation*: Graphics specialists for the production of websites, employing ideas, images and symbols for the design of imagined materializations for virtual space.
3    *Space production*: Virtual visual space being created only through the construction of websites or communications platforms, without "undeveloped" virtual space being possible.
4    *Distances or linking among websites*: The degree and ease of linking among websites, and the efficiency of search engines, both of which meant for connecting between users and information, and, thus, expressing an element of distance within the Internet.
5    *Distances of actors to the Internet*: The level of access to desk and carried computers or to smartphones by users, and their physical distance to such devices, resulting in potential digital gaps, by gender, by social sector or along geographical units.
6    *Individual practices*: Flows, transfers and interactions performed by individuals, using geographical language for their virtual mobility and dominated by globalization of virtual reach.

7  *Networking*: The extent and velocity of connections among users over all their possible networking and email platforms, for exchanges of personally created or transmitted information.

8  *Distanciation*: The utmost time-space "stretch" of human action over the Internet, expressed through the global, borderless and scale-free communications over the Internet.

9  *Time-space compression*: Instant global communications without temporal and spatial limits, thus blurring the differences between local, domestic and global reach, as well as the boundaries between day and night and days of the week, hence requiring constant attention.

10  *Time-space constraints:* Flexible time-space resources for Internet users, notably since no time is consumed for movements among virtual points, and potentially, at least, since virtual space offers unlimited packing capacity. Thus, cycles of production and consumption of virtual action space occur continuously.

## Conclusion

In this chapter we delved a bit deeper into the nature of real and virtual spaces, attempting to explore some basic meanings, significances and workings of these two spaces, notably from the perspectives of their constitution as social action spaces. Four perspectives were offered in this chapter: attributes of real and virtual social spaces; experiencing real and virtual social spaces, notably under the assumption that virtual space has turned into action space; social forces in the emergence of virtual action space; and time geography aspects of virtual action space.

It was shown that the rather numerous attributes, interpretations and metaphors suggested for real social space are all relevant for virtual space as well, though sometimes in different ways than originally proposed for real space, and this despite the lack of geographical scales such as city and region in virtual space. The issue of the experiencing of virtual action space was discussed vis-à-vis Lefebvre (1991), and as developed further by Harvey (1989) and Soja (1996), whose notions of space types seemed to be challenged by the emergence of the Internet as an action space. It was argued that virtual action space is a novel entity which somehow differs from real space in the ways its experienced by individuals, being more flexible in both its very construction and in moving through it.

The social meanings of the construction process of virtual action space were highlighted through structuration theory (Giddens 1990), which accentuates the vicious cycle between human agency and social structures. It was argued that technology has served as a mediator beween human needs and actions regarding speed, as reflected and performed in real space, on the one hand, and business motives, on the other, shaping gradually a virtual action space. Distanciation has reached its utmost in virtual action space, whereas time-space compression has increased to unprecedented levels in its use. Finally, virtual action space was assessed by the time-space prism and time-space constraints proposed by Häger-strand (1970, 1973, 1975). It was argued that the time-space prism cannot be applied with the growing use of the universally and constantly available mobile broadband for accessing the Internet, and that several of the time-space action constraints have either changed or have been nullified for human action in virtual space.

The discussions in this chapter tended to present the seemingly maturing virtual action space as an entity by itself, even if its hardware and its access are rooted in real space. However, as Jordan (2009: 182) noted: "what we once called 'virtual' has become all too real, and what was solidly a part of the real world has been overlaid with characteristics we thought of as belonging to the virtual." With the sophistication of both access to virtual action space and actions performed within it, the very status of real and virtual action spaces is still changing and being challenged, as well as the relations beween them. Thus, some new action processes and patterns may emerge in the near and/or far future, possibly bringing about new rounds of reshaping for our activity customs and patterns, as well as the evolving dimensions of the two spaces.

# 3 Internet operations

Following the conceptual and theoretical expositions of the Internet at large, and of it as being an action space in particular, in the previous chapters, we turn in this chapter to discussions of the Internet as an operational entity, anchored in servers and cables in real space and operated by software in both real and virtual spaces. Thus, our discussions will lead us through several themes: the location of the Internet and the flows it generates; major software systems for its operation through surfing, interaction and contents organization (Microsoft; Google and Facebook); major software for its use and operation in smartphones (iOS and Android); Internet service providers (ISPs) and online service providers, and finally gaps and differences in the adoption and use of the Internet among nations and social sectors.

These discussions are meant to present the real and virtual bases of the Internet as an operational system, constituting the visible and invisible operational tools for individual actions over the Internet, and simultaneously their constituting of major components of the Internet economy. Thus, we are not going to discuss the impact of the Internet on real space "traditional" economies (for such discussions see, e.g., Malecki and Moriset 2008; Moriset and Malecki 2009). Another sector of the Internet economy, namely e-businesses, will be left for Chapter 7, exploring it mainly from users' perspectives, as actors in virtual space.

## Where is the Internet located?

The numerous components of Internet geography have been presented in detail elsewhere (e.g., Kellerman 2002; Malecki and Moriset 2008), so that our focus here will be rather on geographical trends and transitions affecting its leading components in recent years. The locational dimension of the Internet consists of several components: website registration and website hosting services; transmission systems and transmission channels for Internet information; and the dominance of the US in the global operations of the Internet. We will take now a brief look at each of these aspects, moving from production through transmission to consumption.

The geographical pattern of the registration of domains in 2013 has been still typified by a heavy American dominance, with some 59.7 percent of the domains worldwide registered in the US, followed by Germany and China with merely 4.9 percent and 4.6 percent in each country, respectively (Webhosting.info 2013). The suffix ".com" has turned over the years into a dominant suffix code for commercial Internet addresses worldwide, presenting American registrations for website addresses of commercial companies, even when their physical location is in effect in other countries (see Wilson *et al.* 2013).

The supply side of Internet operations in general is characterized by a predominance of the US, and by the concentration of Internet services and infrastructures in it, and we will return to this point again in a later paragraph. Thus, in 2012, out of the top one million websites, 42 percent were hosted in hosting servers within the US, the country that can still provide for smooth access to them from all over the world, given the intensive telecommunications system from the US to other parts of the world. The gap between the US and other countries in this regard is still tremendous, with Germany (7 percent), China and the UK following (3.5 percent in each) in the shares of websites hosted in them (Royal Pingdom 2013). By the same token, the *web hosting services* industry is mostly American, with the US having some 17,999 web hosting companies located in it in 2013, followed by the UK and Germany, lagging far behind with 2,297 and 2,236 companies respectively, and China ranking 16 at that time with merely 347 hosting companies. The centralization of this branch of the real economy of the Internet is even more remarkable within the US, with the company wildwestdomains.com serving some 34.4 million domains, or 25.7 percent of the global market in 2013 (Webhosting.info 2013)!

Geographically, the preferred state within the US for the web hosting and data center industries is Oregon, in northwestern US, serving both large computer companies, such as Google and Facebook, as well as companies offering hosting services for the websites of numerous smaller companies ("co-location") (Blum 2012; *The Economist* 2012). Oregon has become attractive for this industry for several reasons: its proximity to the large market of California, measured by the minimal flow time of large volumes of data; the convenient access to cables to Asia; low taxing; the use of the dry desert air for the heavy cooling systems required for server farms; ample supply of electric power for both the operation and the cooling of servers; available manpower; and the cumulative clustering effect which has drawn additional companies to Oregon over the years.

Following its very production, registration and hosting, a website is ready for access by users through the Internet transmission system. This system may be viewed as consisting of four major physical layers: the *physical layer; data link; network layer* and *transport* (Gorman and Malecki 2001). Thus, a session of Internet use may generally look like this, in a mostly simplified way: a user approaches the system through the physical layer which constitutes the web of lines connecting computers worldwide, as well as the communications systems in general. When a call is made from an end computer by an Internet user to another computer, mostly a server hosting a website, and normally executed through the user's *Internet Service Provider (ISP)*, the data link checks the unique numerical address of the called computer-server and its location on the network. Then, the network link establishes the best path to move data between the calling and called computers through the use of routers. An interactive session or reliable transport of data is then established. Similarly, connections are established between ISP computers and website hosting servers.

We will now turn our attention to the global system of channels providing for the flows of Internet traffic. Cross-country, international and intercontinental transmission of Internet information is channeled through "Internet backbones," which serve as "highways" for Internet traffic. These are typified, like transportation highways, by their speed, accessibility and connectivity. Again like transportation routes, their crucial role is to facilitate the efficient flows of the Internet products, namely information and knowledge, between producers (websites) and consumers (Internet users) (O'Kelly and Grubesic 2002), as well as among consumers themselves via email. Once again as with transportation networks, it is not always clear whether demand (information) brings about supply (backbones), or the other way around, and in a dynamically developing system such as the Internet has been, demand and supply may also emerge simultaneously. Internet backbones may constitute in most cases independent fiber-optic lines, and in other cases they may constitute leased telephone lines.

The international flows of Internet traffic are channeled through systems of satellites, mainly American and Russian ones, and the much more popular system of submarine cables. The geography of satellites, which are used mainly for TV transmissions and also for some telephone traffic, has been presented elsewhere (see, e.g., Kellerman 1993; Warf 2013), so that our brief discussion here will rather focus on the submarine cable system.

The international map of submarine Internet cables presents a well-interconnected world, coping recently with a rather growing demand for heavy video transmissions (Kellerman 2010) (Figure 3.1). As a matter of fact, some 20 percent of the existing submarine cables are not in lit (used) anymore, as newer and faster cables have been put into service (Malecki and Wei 2009). The transatlantic Internet transmission system has been saturated as of 2003 and no new cables have been added there since then, whereas the growing demand for Internet traffic in Asia has brought about recent additional cables. Still though, the trans-Pacific capacity is lower than that of the trans-Atlantic one, and trans-Russian continental cables have introduced a competitive alternative to submarine trans-Pacific cables (Malecki and Wei 2009). Some additional new cables serve already and will be put into service in the near future for growing demands in Africa, Latin America and the Middle East (Oremus 2013).

The global Internet cable system is interconnected through Internet exchanges, and at the time of this writing there were some 300 such exchanges active worldwide (Telegeography 2013b), many of which located in major cities, which, on their part, present high demand for Internet traffic, so that these exchange-hosting cities reflect traditional urban hierarchies (Malecki 2002; Tranos and Gillespie 2011b). In Europe, the major Internet hubs constitute the so-called *Internet diamond* cities consisting of London, Paris, Frankfurt and Amsterdam, representing economic and financial centrality for the first three

*Figure 3.1* Submarine Internet cables, 2013 (source: Based on Telegeography 2013a).

cities, and historical international transportation and trade centrality for Amsterdam (Kellerman 2002). These four cities serve further as the major aviation hubs for Europe, so that human physical and virtual mobilites are interconnected (Tranos 2011a; Kellerman 2012a). The very landing posts for submarine cables have been selected given several factors: favorable offshore costs as in Mumbai (Malecki 2009); convenient continental access for such cables, such as in Palermo; (Sicily, Italy) (Tranos and Gillespie 2011b); and in Fortaleza (Brazil) (Oremus 2013); or because of relatively more favorable political conditions (e.g., Djibouti) (Oremus 2013).

We mentioned already the American predominance in the Internet operations system with regard to website registration and hosting, as well as concerning its leading international connectivity for the transmission of Internet information. Let us now take another look at the American status vis-à-vis the Internet, turning our attention now to users and usage. We will see that the US has lost some of its leading status with regard to its number of users, but it is still the leading country as far as the routine access of the Internet by users worldwide.

We noted in Chapter 1 that the Internet was developed at the time in the US, and was put into open service first in the US. Until the mid-2000s, the demand side of the Internet, namely the spread of its users worldwide, has reflected an American dominance, since the US was the country with the largest number of Internet users, though it was only the third largest country worldwide by its population size, with China and India each about four times larger. This is not the case anymore: as of the mid-2000s the number of Chinese Internet users has surpassed dramatically that of the US, reaching some 389 millions in 2009 as compared to 245 million American ones by then, and with India still lagging enormously behind with just 61 million users (CIA 2013). However, despite the impressive number of Chinese Internet users, the proportions of the population in the US and China using the Internet presents a wide gap between the two countries: 77.9 percent in the US in 2011, and just half of it, 38.3 percent, in China at that time (ITU 2012). It is not for the US, though, to lead the globe in the shares of populations using the Internet, this is rather left for the Nordic countries, which lead with some 90 percent of their populations using the Internet. We will further address international and inter-sector gaps in Internet use in a later section in this chapter.

The virtual space of the Web extends virtually between the physical locations of users and those of website hosting servers, connected through the cable and exchange systems. In terms of the spatial pattern of real world Internet traffic, over one-third of the global population accessing the Internet from anywhere on earth (and over two-thirds in developed countries) would probably reach the US in some way for surfing or using the Web for any desired action: there is a 60 percent chance that the website they approached is registered there; some 42 percent chance that the website is hosted there, and a high chance that the connection to the website they approached, even if the website is domestic, was routed through the US. Since there are no data on flows through the Internet backbone system, one cannot fully estimate the percentage flow of Internet traffic through the US. Warf (2013: 16) estimated it at 75 percent of global Internet traffic! This condition of a geographically spreading demand or users coupled with a still US-centered supply gives provides the US with an especially strong power in the Internet system. However, this power seems mostly economic, since a potential shutting down of the system by the US would not abort the system completely, as the flow system would still be at work through the use of automatic routers diverting traffic to alternative routes, and many or most websites would still be accessible through temporary locations, employing peering (peer-to-peer, P2P), and other non-US based technologies and systems. Such alternative flows, selected by routers, came into effect automatically, at the time, following the 9/11 2001 terrorist attack in New York, which put out of service a major Internet hotel (server farm) located next to the destroyed twin towers.

## Operational software for the Internet

In their book on the relationships between space and code, Kitchin and Dodge (2011: 71) state that, "code/space is quite literally constituted through software-mediated practices, wherein code is essential to the form, function and meaning of space." These crucial roles of software which were originally noted for real space are even more striking for virtual space, the very existence and functioning of which is fully dependent on software. Thus, in this section we will focus on operational software for the Internet at large, followed in the next section by a discussion of software dedicated for the operation of smartphones. General operational software is a must for any uses of virtual space by individuals and for their diversified operations within it. Similarly to the American dominance in the operations of the basic real space dimensions of the Internet, here too the major commercial players in the market of general Internet software are American, even if the relevant companies have turned over the years into multinational and public ones.

The term *operating system*s has been assigned to software packages which provide an interface between hardware components (such as computers, disks, etc.), on the one hand, and dedicated software for work with computers and their management, on the other. Operating systems also provide for connections and interactions among these software components. With early roots in the 1950s, operating systems for individual users emerged as of the introduction of stand-alone personal computers (PCs) in the mid-1970s, in parallel to the early development of information and communications systems for individual users. The development of operating system packages matured with the introduction of Windows by Microsoft in 1985. Microsoft also used the idea of software packaging for its Microsoft's Office package, which was introduced in 1989, consisting of several components, such as word processing, spreadsheet software, slide construction and presentation, etc. Thus, when the Internet was introduced in its current form in 1994, the basic enveloping, access and usage tools for it were there already, except for browsers. The first popular browser at the time was Netscape, and Microsoft introduced in 1995 its Microsoft Network (MSN) package as an umbrella for a variety of information and communication services for individual users of the Internet (including email; chatting; video messaging; news service, etc.). Side-by-side, Microsoft introduced its Explorer browser, which has soon inherited Netscape as the most popular browser.

Over the years, with several competitors emerging, the status of Windows as an operating software for PCs and laptops has remained strong, and in 2013 it enjoyed an 85.5 percent market share (Stat Owl 2013a). This market dominance has been even more striking with regard to its Office package which enjoyed a market share of 95 percent in 2013 (Forbes 2013). However, Google Doc now challenges this market, emerging as a major competitor for Microsoft at large, and notably regarding browsing. Thus, Explorer, though still the most popular browser, enjoyed in 2013 a market share of only 43.9 percent, followed by Google's Chrome with 24.5 percent (Stat Owl 2013b).

Whereas Microsoft was founded and has become a significant operation system producer already before the maturing of the Internet, Google came into being as an Internet-dedicated company. Established back in 1996 and launching its search engine a year later, its early business concentrated in Web search, or in the organization of Internet information. As of 2003 Google has moved also into areas of information production, such as citation and information ranking, satellite and street images, digital book images, online translation and the online posting of video clips. The development of these products required Google to establish its own R&D (research and development) centers, and these centers have further developed its Android operation system for smartphones, thus turning Google into "a major gatekeeper for digital information" (Paradiso 2011: 54). In 2013, Google was the leading Web search engine service enjoying some 80 percent of the global market (Stat Owl 2013c).

The third major Internet operational software, Facebook, was established much later, as compared to Microsoft and Google, following the introduction of Web 2.0, back in 2004. As a networking tool the use of Facebook requires free subscription, and the three major software tools described so far differ in this point from each other. Microsoft software, mainly Windows and Office, have to be bought and licensed, Google provides free access to its services, whereas Facebook requires free subscription. Despite its young age, Facebook has turned into the most popular social networking tool at an extremely fast pace. Having its roots among students in leading American universities, Facebook has reached almost one billion subscribers worldwide in 2013 (Checkfacebook 2013), amounting to some 43.5 percent of Internet users worldwide at that time. Facebook permits the posting of people's identity information, as well as their blogs, their reactions to posts on fellows' blogs, and more generally providing for wide interactions among its subscribers, including video calls. It further permits the installation of websites. Its availability on smartphones has meant for many of its subscribers a most frequent use of it (Kellerman 2010).

## Operation systems for smartphones

As we noted in Chapter 1, smartphones were introduced back in 1993, permitting communications through Wi-Fi. However, they have become widely adopted as of the 2000s, when access to the Internet has become possible through them. The installation of Internet access components in smartphones required the development of mobile operation systems specifically dedicated to smartphones (coupled later on also with such systems for tablets), and these operation systems include also a browser, and they further provide for screen touch options. These operation systems for smartphones have attempted to present users with a use experience similar to that of Internet use on PCs, thus turning the very use of the Internet completely mobile.

Nokia's Symbian, commercially introduced in 2000, was the first operation system for smartphones, side-by-side with several other systems that were introduced by Microsoft also as of 2000 (e.g., Pocket PC 2000, 2002). The two currently most popular systems were introduced later on: iOS for Apple's iPhone and Google's Android for numerous other ones (e.g., Samsung's Galaxy, Nexus, LG) were introduced only in 2008. In 2012 Microsoft's Windows Phone 8 replaced Symbian in some of Nokia's smartphones. In the last quarter of 2012, some 51.5 percent of the sold smartphones worldwide were Android-based, as compared to 23.6 percent for iOS, which is only installed in Apple's iPhones. Microsoft's share reached a mere 1.8 percent at that time (Gartner 2013). This market share has marked a dominance of Google in the mobile phone operation systems market, as compared to that of Microsoft for PCs.

## Internet service providers

Internet service providers (ISPs) are companies which operate the transmissions or the flows over the Internet, either those which do so along the Internet backbones, or those which provide local Internet access and content to customers' end computers. ISPs have been widely discussed elsewhere (e.g., Kellerman 2002), and it will suffice here to note on the structure of this service. *Transit backbone ISPs*, or *Tier 1 networks* (T1), own and manage backbone transit, so that all other lower-level transmission services depend on them (e.g., AT&T, UUNet). The National Science Foundation established the first terrestrial backbone ISP across the US back in 1987.

Numerous lower level ISPs (T2-T3) depend on the backbones, and they include: *downstream ISPs*, namely the local and regional ISPs; *Web hosting companies* specializing in the owning and managing of server farms for website hosting, and which we reviewed in a previous section; and *online service providers* (OSPs), which provide for customer-visualized access and interaction with Internet content. Frequently, local ISPs serve also as OSPs, thus providing end customers with both Internet transmission and content services, for example AOL, Verizon and AT&T (US), Nippon (Japan), and Telecom Italia (Italy). Another classification of ISPs, international ISPs (IISPs), focuses on the geographical range of services, thus distinguishing among global, regional, national and academic ones.

## International differences in Internet use

International differences in Internet use can be approached from three perspectives: digital gaps among countries in the national rates of access to the Internet; differences in government-forced restrictions or censorships over Internet resources; and cultural differences among countries in the adoption and use of the Internet. We will briefly review each of these categories of international differences, whereas Warf (2013) and Wilson *et al.* (2013) present recent detailed geographies of the Internet by world regions.

Table 3.1 presents the 2013 global digital gaps in the penetration rates of the Internet by world regions (see also Appendix for country specific Internet data 2000–2012). The data on penetration rates, provided by the International Telecommunication Union (ITU), do not differentiate, though, between North America and Latin America, thus concealing an additional digital gap on the global map, that in the Americas. In some way, the classes of Asia and the Pacific, as well as that of the Arab States, lump together countries with differing levels of exposure to the Internet, such as globally leading Japan and Korea, on the one hand, as compared to Vietnam, on the other. The data reveal a gap in the use of the Internet largely corresponding with the more general gaps in the levels of economic development, with Africa being the most lagging continent and Europe the leading one, at the ratio of 1 : 4.5 between these two continents, as far as the population shares of Internet users is concerned.

Gaps are even more striking with regard to the more advanced and more expensive mobile broadband subscriptions reaching the ratio of 1 : 6.2 between Africa and Europe. On the other hand, however, the data on mobile phone subscription show that most world regions reached some saturation in this regard, and even in Africa close to two-thirds of the population subscribed to mobile

*Table 3.1* Global digital gaps: world regions by penetration rates 2013

| World region | Internet users | Mobile broadband subscribers | Mobile phone subscribers |
|---|---|---|---|
| Africa | 16.3 | 10.9 | 63.5 |
| Arab States | 37.6 | 18.9 | 105.1 |
| Asia and Pacific | 31.9 | 22.4 | 88.7 |
| Commonwealth of Independent States (CIS) | 51.9 | 46.0 | 169.8 |
| Europe | 74.7 | 67.5 | 126.5 |
| The Americas | 60.8 | 48.0 | 109.4 |
| Developed countries | 76.8 | 74.8 | 128.2 |
| Developing countries | 30.7 | 19.8 | 89.4 |
| World | 38.8 | 29.5 | 96.2 |

Data source: ITU 2013.

telephony in 2013. This high penetration of mobile phones in the developing parts of the world, leapfrogging fixed telephony and fixed Internet connection, may possibly bring about in the future a wider penetration of mobile broadband connection to the Internet in developing countries, possibly occurring once both the devices and their use will become cheaper, as has happened already so far with mobile telephones and their adoption. However, such possible wider penetration of the Internet might be restricted by levels of literacy, which is a must for the use of most Internet resources other than music, films and clips.

Governmental policies may play an important and sometimes crucial role in the development and routine use of the Internet for both better and worse (Farivar 2011). Thus, the major financial and other investments by the South Korean government along decades in Internet infrastructures have yielded the most advanced country as far as broadband speeds are concerned, permitting users to easily transmit heavy files, consisting mainly of films and music. At the opposite end, blogging in Iran, which assisted political unrest at the time, has been banned, jointly with the blocking of YouTube and much of the Web content, which is filtered; this is coupled with arrests and expulsions as punishments for the free use of the Web. China has been involved in similar steps, also on cultural grounds (Lagerkvist 2008), whereas other countries have attempted to censor specific items for various religious and political reasons (*The Economist* 2012). We will return to governmental censorships later on in Chapter 9.

There might, however, evolve some domestic circumstances, habits and cultures, which may turn the use of the Internet in some ways into a specific local experience, thus bringing about the emergence of numerous Internet cultures. For instance, the advance payment by Internet customers to online companies via credit cards, which has become the routine procedure in Western countries, is not the case in Russia, China and India, where customers prefer to pay for merchandise purchased online only upon its delivery. The very use of the Internet may further be restricted in time by its cost, when modems, other than those used for fixed broadband, are used for communications and the Internet use being charged by time, as is the case in many African countries (*The Economist* 2012).

In a comparative study of 45 countries using data for the late 1990s, Park (2001) claimed that the adoption of the Internet and its use have been heavily influenced by cultural aspects. Generally, the level of penetration of the Internet has depended on the social distance of countries from the US, which is the country which has been most reflected in the Internet. Thus, it was difficult at the time for Japanese to adopt the Internet, which permits rather "horizontal" and open interactions by its users, without regard to hierarchical social structuring, in a rather "vertically" highly hierarchically structured society. Hence, Shiu and Dawson (2004) stated that the level of adoption of emailing may differ from the adoption of online shopping, which on its part might be attractive when the time budget of individuals becomes restrained, thus making technology overcome cultural-national values. These two internationally comparative studies further claimed that a dominance of feminist and individualist values in countries may assist the adoption of the Internet, as compared to countries with prevailing collectivist and uncertainty avoidance values, since emailing permits free networking which may be preferred by women.

Time is needed in order to overcome habits and cultural traditions, and this may be accentuated by looking at the Internet experience of the US, the country which adopted e-commerce and e-learning first, and the country which still dominates the e-commerce scene (as we will see later on for both online shopping and distance learning in Chapter 7). The initial advantage of the US at the time, as the country in which the Internet was invented and first widely adopted, was that the cultural ground was prepared for wide and rather advanced e-commerce and e-learning uses, even when the US was still lagging behind Korea, at the time, in fixed broadband penetration. Thus, back in 2004 some 50.9 percent of total global e-commerce took place in North America (Hwang *et al.* 2006), but back in 2000–2001 only some 56 percent of the degree-granting institutions in the US offered any distance courses, but not necessarily online full-degree studies (US Bureau of the Census 2006), a trend which has changed later on (see Chapter 7). In the currently emerging mobile broadband age, it was reported for 2009 that 21 percent of young US customers used mobile phones for banking and that 25 percent of mobile phone users in general shopped online (Scherr Technology 2009).

In Japan, in which email communications were foreign to its social culture of a rather accentuated hierarchical order, by the early 2000s customers adopted Internet browsing and e-commerce over mobile phones, at a time when such applications were in their infancy elsewhere (Kellerman 2006b). Aoyama (2003) attributed this pattern to a variety of specific social conditions, ranging from market positioning of m-commerce (mobile commerce), through the importance of portability and urban spatial structure, to socially-embedded user friendliness attributed to m-commerce (see also Zook *et al.* 2004).

## Societal differences in Internet use

The commonly used terms "digital gap" and "digital divide" seem to carry the connotation of "haves" and "have-nots," with regard to the very availability and adoption of the Internet and related devices by regions or countries. Gilbert *et al.* (2008) suggested the widening of the scope of digital gaps by looking at numerous societal contexts of the availabilities and uses of ITs. Thus, they proposed to examine the delivery of information through ITs; the settings in which technology is accessed; the role of social networks in the shaping of access and use of technologies, and, finally, social policies regulating access to technologies.

The emphasis on the dimension of the very access to the Internet leaves in the shadows yet another important dimension, namely that of differences among social sectors, as far as the levels of use of the Internet by those who have gained access to the Internet already. More specifically, it is important to examine some possible differences in the level of Internet use, or the degree of emergence of a second and rather virtual action space, along gender, age, social sectors and geographical location (center/periphery), in an attempt to see whether the frequently observed societal gaps in people's exposure to technologies, employment patterns, legal rights, etc. among demographic and social sectors, would be relevant to the use of the Internet, as well. In other words, the question is whether women, children and adolescents, and residents of peripheral locations who already enjoy access to the Internet would make lower use of it, as compared to other sectors. As Warf (2013) recently noted, data on differing levels of Internet use are scant, but as we will see in the following paragraphs some of them might be of interest.

It was estimated for 2013 that, globally, women lagged behind men in the level of Internet adoption, with a gap of 25 percent (reaching some 29 percent in Europe) (Brain Track 2013), but for the US an equality was shown to exist by then, with 85 percent of men using the Internet, as compared to 84 percent of women (PewInternet 2013c). The ITU (International Telecommunications Union) reported in 2013 of only three countries in which the percentage of women using the Internet surpassed that of men, even if only marginally: Ireland (80 percent for women versus 78.6 percent for men in 2012); Jamaica (29.8 percent for women versus 25.4 percent for men in 2010); and closely so in Thailand (23.4 percent for women versus 23.9 percent for men in 2011). However, women's wider exposure to the Internet may emerge within regions, and even in peripheral areas, such as in rural Russia, where women outnumber men in their connection to the Internet, given problems of unemployment and alcoholism among the latter (Dovbysh 2013).

As far as gender differences in the uses of the Internet among those connected to it are concerned, one may assume no differences in the professional arena, as was shown, for example, for practitioners in the public relations business (Porter and Sallot 2003). Women, though, were found, at the time, to be less represented in the business of IT design (Fountain 2000). However, there are gender differences in the personal uses of the system. Thus, it was shown, for instance, that women's uses of the Internet for maintenance activities, such as for bill paying and reservation making, substitute physical travel, whereas men's leisure uses of the Internet reduce their physical travel for these purposes (Ren and Kwan 2009).

Women's traditional higher tendency for networking has been expressed in data on Facebook subscription by gender breakdown. Thus, globally it was assumed for 2013 that 53 percent of the subscribers were females (Zephoria Inc. 2013). For the US it was shown that 72 percent of female Internet users were Facebook users in 2012, as compared to just 62 percent of males (Duggan and Brenner 2013). More female Facebook subscribers were also observed in Brazil for 2013, whereas in India only 25 percent of Facebook users were women, and in Mexico and Indonesia there were slightly more men than women among Facebook users (About.com 2013). These differences probably attest to wider differences among countries in women's access to the Internet in general. Men and women may differ also in their preferred Internet contents. It was found for American Asian-Indians that men were more interested with news, sports and political events in India, as well as in family ties, whereas women preferred to use the Internet for the shopping of ethnic products and for accessing elements of traditional culture (Adams and Skop 2008). Similarly, a survey in Finland in 2011 found that enhanced use of the Internet among women implies higher use of e-government services, something that is not matched by men (Sakari 2013).

When it comes to the age of Internet users, it seems that the Internet in general and social networking in particular belong, foremost, to younger people. Unfortunately, data on the global age breakdown of Internet and social networking users have not been available, but Table 3.2, presenting data on the US in 2012–2013 may hint to similar distributions in other countries as well. The 95 percentage of Internet users among teens has been steady since 2006 and is has continued into 2013 as well (Madden *et al.* 2013). Furthermore, a 2012 survey among adolescents aged 12–17 years in five leading cities in East Asia (Hong Kong, Seoul, Singapore, Taipei and Tokyo) revealed that 90 percent of them used the Internet (Lin *et al.* 2013).

The higher Facebook percentages than social networking ones in some of the classes of Table 3.2 may reflect Internet users who make use of Facebook, but not for networking. The trends in all the three categories of Internet users, social networking users and Facebook users are the same: high percentages among young people in general; peaking among the youngest adults (18–29) and declining among older adults along age. Furthermore, older users may tend to use the Internet fewer hours than young ones (Adams and Skop 2008). In some cases, though, the young ones may use the Internet also for the older ones, such as teens seeking health information over the Internet for their low education parents (Zhao 2009).

The contemporary scenario of decline in the uses of the Internet with increasing age reflects even now the newness of the Internet, introduced in 1994, and even more so of Facebook, introduced only in 2004. In decades to come, one can assume more equal levels of Internet use along the age strata and lower declines in social networking by age, since the current numerous young users will be the future older ones. The Internet has an advantage when compared to the equivalent technology of personal corporeal mobility, the automobile, since the driving of automobiles requires licensing and this becomes available only

*Table 3.2* Percentage Internet and social networking users by age in the US 2012–2013

| Age group | Percentage Internet users 2013 | Percentage social networking users among Internet users 2012 | Percentage Facebook users among Internet users 2012 |
| --- | --- | --- | --- |
| 12–17 | 95 | 81 | 77 |
| 18–29 | 98 | 83 | 86 |
| 30–49 | 92 | 77 | 73 |
| 50–64 | 83 | 52 | 57 |
| 65+ | 56 | 32 | 35 |

Data sources: PewInternet (2013c); Duggan and Brenner (2013); Madden *et al.* (2013); Brenner (2012).

*Table 3.3* Percentage Internet and social networking users by location type in the US 2012–2013

| Location type | Percentage Internet users 2013 | Percentage social networking users among Internet users 2012 | Percentage Facebook users among Internet users 2012 |
|---|---|---|---|
| Urban | 86 | 70 | 72 |
| Suburban | 86 | 67 | 65 |
| Rural | 80 | 61 | 63 |

Data sources: PewInternet (2013c); Duggan and Brenner (2013).

between the ages of 16–18, depending on country. The Internet, on the other hand, requires only literacy, and is, thus, available as of early ages, assuming that its access is reachable (see also Kellerman 2006).

A third societal differentiation in Internet usage is location, differentiating among urban and rural regions. Thus, Table 3.3 presents percentage Internet and social networking users by location type in the US in 2012–2013. The higher Facebook percentages than social networking ones in some of the classes may reflect Internet users who make use of Facebook, but not for networking. As has been argued time and again for the US, the lower rural levels in the use of the Internet as compared to those of urban and suburban areas reflect problems of human capital, mainly lower incomes and education levels, rather than lower levels of Internet infrastructures (Mills and Whitacre 2003; Basu and Chakraborty 2011; Malecki 2003). This was shown for other countries, as well, e.g., China (Guo and Chen 2011) and Italy (De Blasio 2008), and in it was further argued for the latter distance from major urban areas still counts.

One would have expected, though, to find higher levels of networking in rural regions, given their peripheral locations. The lower suburban and rural levels of networking in the US, as presented in Table 3.3, possibly attest to preferred live face-to-face social networking within livelier social communities. A study in the Russian periphery suggests that the use of the Internet is more restricted there, since residents there consider Internet information as unreliable (Dovbysh 2013). The digital divide between urban and rural areas has drawn the attention of numerous governments worldwide, leading them to activate development plans for its reduction. This has been the case, for example, for India (James 2003); Mongolia (Ariunaa 2006); Hungary (Pósfai and Féjer 2008) and Chile (Kline 2013).

## Conclusion

We are now able to portray some of the features for the average structure of the Internet action space for individual users of the Internet. Such users comprised in 2013 some three quarters of the population in developed countries and about one-third of the population in developing countries. All in all, there are two leading trends which characterize the Internet at large and its being an action space in particular: the dominance of the US and the dominance of major commercial players. This dominance brings about a rather global similarity of the technical, visual, and operational structures and envelopes of the Internet, so that when a computer is used by foreign visitors in some country, not being able to read the domestic language, they can still navigate the system, given the universal design of the screens and virtual buttons. Side-by-side with this global operational similarity, the very ways of use of the Internet, as well as the preferred applications may depend on domestic cultures. However, such domestic cultures may, with growing digital globalization, present some diminishing impacts in years to come.

With 42 percent of the websites hosted in the US and almost 60 percent of them registered there, there is a good chance that about one half of individuals' uses of websites have some American component in it. Moreover, even if a domestically registered and hosted website is accessed by users there is some chance that this inland connection between users and websites may be actually pass through the US, which is the most widely connected country, through both trans-Pacific and transatlantic cables. It is almost ubiquitous that the enveloping and major operational tools used by individual Internet users are once again American and universally adopted all over the world, such as Windows, Word and Google. Similarly there have emerged dominance patterns in smartphone operational systems with the wide adoption of Android and iOS. It is only for browsers to present a slightly wider market beyond the two major players Explorer (Microsoft) and Chrome (Google). There is further a high chance that users worldwide may belong to geographically wide-ranging networking system, either Facebook, which may soon include one half of global Internet users, or other popular networks, such as Twitter.

This scene of almost full global standardization of the technical, visual, operational structures and envelopes for the Internet, reminds one of the urban landscapes of real space cities. These too have similar elements worldwide, consisting of major roads and highways, coupled with more minor streets stretching out of them; high and low buildings; stores at the street levels of buildings; street lights, etc. These similarities in the general components and structures of urban landscapes permit the use of the same digital navigation systems (GPS – Global Positioning Systems) globally. By the same token, website developers need to care only for language choice (or direct users to the proper Google translation tools), so that the Internet action space and its ways of operation and use have continuously moved towards universal standardization with regard to both their operation and visualization. Furthermore, Internet users can make use of the system using computers in any country without knowledge of the domestic languages, since the appearance and structures of Windows and Google screens are similar worldwide.

The societal breakdown of Internet users and uses seems to be quite complex. Whereas the exposure of women to the Internet is lower than that of men in most countries, either marginally or significantly, their preferred personal uses of the system differ from those of men, with an accent on networking. As compared to other technologies, the Internet is a technology which has been widely adopted by children and adolescents, and even more so by young adults, again with a high popularity of networking. However, the penetration of the Internet by geographic regions points to higher levels of adoption in urban areas, given the differences in human capital between urban and rural areas.

# Part II

# Human needs and the Internet

# 4 Human basic needs and their provision

In the first part of the book we attempted to present the numerous dimensions of virtual space as an arena for human action, second to real space: the very notion of virtual space, notably as compared to real one; the nature of the Internet; theoretical and conceptual aspects of virtual space as social space; and, finally, the operations of virtual space through the Internet. Following this exposure of the platform of virtual space and the Internet, in this second part of the book we will become acquainted with three rather primary human needs and with their possible provision through the Internet: basic human needs; curiosity and the expression of identity.

## An overview of basic needs and space

Virtual space will be presented in this chapter as a space through which basic human needs can be gratified, alongside with the satisfaction of basic human needs which has been traditionally attained through real space (Kellerman 2013). It will be shown that the role of virtual space in this regard, as compared to that of physical space, becomes more significant along the common hierarchy of human needs: physiology; safety; love/belonging; esteem and self-actualization. This growing role of virtual space has brought about an equivalent hierarchical relationship of virtual space with physical space: complementarity; competition; substitution; merger; escape and potentially also exclusivity. The new role of virtual space as facilitating the satisfaction of basic human needs, and the potential relations between the two spaces are interrelated, since if we assume that virtual space can satisfy basic human needs then this may imply some kind(s) of relationships with physical space as the past sole provider for the satisfaction of these needs. It will be further shown in this chapter that escape from physical space to virtual space, in the constitution of escape as both a need and a relationship, has been brought about by social networking. Networking is similar in this regard to activities of escape within physical space, mainly as offered by tourism. Finally, it will be argued that it does not seem real to foresee that virtual space will offer exclusive fulfillment of, as of yet unforeseen, new human needs, needs which will not find satisfaction in real space.

As we noted already in the first part of the book, in less than two decades the Internet has been formed and shaped within virtual space or cyberspace as a space containing facilities which permit activities which prior to its inception were reserved for sole existence and operation in physical space, mainly within urban contexts. Side-by-side with the emergence of the Internet, fixed and mobile telephone services have been digitized and widely adopted, thus providing for additional virtual arenas for their subscribers. Following its maturity in the 2000s, the Internet has usually been assessed as a space of human activity consisting of widely varied categories and forms, such as written, voice and video communications, as well as numerous e-functions (e-commerce; e-government; e-learning, etc.), which have traditionally been offered in physical space only (see, e.g., Loo 2012; Warf 2013; and Chapter 7).

The growing adoption, dissemination and use of the Internet and of virtual space in general in recent years, call for their assessment also from yet another, and more basic perspective, namely their role as facilitating the very satisfaction of human basic needs and motivations, and we will focus on this in this chapter. Such facilitation has become possible through the means of the numerous virtual activities made available through virtual space. In other words, the satisfaction of basic needs is a must for humans, and some of this may now be achieved through virtual communications and services. As such, virtual space may have come to form in recent years, jointly with physical space, a "double space" for the satisfaction of human needs in the two spaces, rather than the traditional single physical space as the only arena permitting this gratification.

As we noted already, a vast majority of contemporary individuals in developed countries are continuously exposed to the Internet and to its numerous uses, though the digital divide among population sectors and regions may still prevail there, similarly to the international gaps that we have seen in previous chapters. Subscribers to the Internet and to mobile phones, notably smartphones, tend to function daily and continuously within a "double space" – the physical and the virtual. This human functioning in two spaces, and the very simultaneous human presence in the two spaces, has routinely been termed as co-presence (see, e.g., Urry 2002; Kaufmann 2002). However, as things currently seem to be maturing from a virtual space into a space that permits a wide range of operations within it, this makes the relationships between the two spaces even more complex. Users tend to perform daily activities in combined ways between the two spaces, such as touching and trying products in real stores and then purchasing them online or vice versa (see, e.g., Schwanen *et al.* 2008; Dodge and Kitchin 2001; Kellerman 2010). Thus, the use of virtual space does not only permit mere human presence in it simultaneously with presence in real space, as it may also, potentially at least, provide for a number of relations with physical space (competition; complementarity; substitution; escape; merger and theoretically at least, also exclusivity).

Our exposure here of human basic needs and their satisfaction will be highlighted through Maslow's (1943, 1954, 1968) theory, which we will briefly present in the following section, and which has not received any attention in the geographical literature. Geographers could have attempted to highlight the spatial perspectives of human basic needs, in the sense of identifying the locations for the satisfaction of these needs. This lacuna in the study of the spatial dimension of the satisfaction of basic human needs has persisted despite the growing tendency by geographers over the years to tackle psychological issues. The lack of interest by geographers in the satisfaction of human basic needs might have been related to the specific locations of facilities in real space dedicated for the gratification of human needs, such as homes and work places, which have not posed a challenging interest for geographical inquiry from the perspective of basic human needs, rather being studied through residential and employment perspectives. The lack of geographical attention to human basic needs and their satisfaction might further be related to the view of "space" as a container of facilities for the sake of the analysis of satisfaction of human needs, rather than its being viewed as a relation, as we noted in Chapter 2 regarding social space (see, e.g., Lefebvre 1991). However, under the contemporary conditions of two spaces becoming available for human activity, this topic deserves geographical attention, in an attempt to assess the possible gratification of human needs over the Internet, as compared to the traditionally existing possibilities in real space. Furthermore, and from a more general perspective, it is of importance to highlight some of the limits for human activity over the Internet, as compared to activities in real space, notably at times when the Internet seems to constitute a limitless spatial entity.

We will begin our discussion in this chapter with a brief exposition of human needs and their gratification, attempting then to assess the roles of the two spaces in this regard. This discussion will then be followed by an examination of the numerous emerging relationships between the two spaces from the perspective of their provision for the satisfaction of human needs. Finally, we will explore the emerging trend of the Internet possibly becoming a need by itself, side-by-side with its being a platform for the satisfaction of other needs.

## Human needs

The exposition of human needs has constituted one of the foundations of modern psychology, based originally on Maslow (1943, 1954, 1968), who proposed a five-phase hierarchy of these needs, presented later as a pyramid of human needs (Figure 4.1): physiological needs; safety; love/belonging; esteem and self-actualization. Though widely accepted as a tool for the understanding of classes and levels of human needs, Maslow's hierarchy of needs has been most widely discussed, criticized and challenged over the many years since its first presentation (for reviews and criticism of Maslow's original hierarchy see, e.g., Neher 1991; Kreitner and Kinicki 2008; Myers 2009). Critics have mainly questioned the following aspects of the hierarchy:

1  The specific needs included in each phase.
2  The very notion of a hierarchy of needs in which only the fulfillment of needs in one phase may permit the rising of higher needs, as opposed to an alternative possibility of simultaneous fulfillments of human needs, and

another option of a lack of satisfaction of lower needs as leading, rather than preventing the fulfillment of higher-level needs.

3   The cyclic nature of the more basic needs, as compared to the more continuous higher ones.
4   The very validity of the assumption that higher needs are innate, thus ignoring the significance of culture and environment.
5   The relevance of the hierarchy to differing cultures.
6   The difficulties in the empirical testing of the theory.

These criticisms have led to several revisions and updates that were proposed over the years for Maslow's hierarchy (see, e.g., Adelfer 1972, 1989; Kenrick *et al.* 2010; Tay and Diener 2011). However, in spite of its limits, the theory "became immensely popular, because it is intuitively logical" (Wade and Tavris 1987: 380), and because "it highlights the complexity of human needs" (Feldman 2003: 255). Thus, the theory "enjoys wide acceptance, especially among humanistic psychologists" (Neher 1991: 89), and "Maslow certainly deserves credit for his general thesis" (Neher 1991: 109).

As we mentioned before, it seems that geography and geographers have tended to ignore Maslow's theory over the many years since its inception, with the exception of tourist geographers whose contribution we will discuss later on. Thus, the set of human needs/demand for basic needs and their supply have not been treated systematically within geography. The list of these needs, as

*Figure 4.1* Hierarchy of human needs (source: Kellerman 2013).

portrayed by Maslow, and basically accepted by most of his criticizers, is, though, of importance when a comparison of their supply through physical and virtual spaces is sought. By the same token, the hierarchical order of these needs as portrayed by Maslow, even if criticized, provides a conceptual order that would suffice for an initial examination of the spatial dimension of the satisfaction of basic human needs, under the assumption that people are active in both real and virtual spaces.

The numerous needs of human beings lead, by their very nature as needs, to motivations for their fulfillments. From a geographical perspective these motivations have brought about the construction of physical facilities, such as buildings, as well as the construction of virtual ones vis-à-vis websites, for the satisfaction of these needs. Such facilities may include homes and fields for the provision of the most basic needs for food and shelter in the lowest phase of physiological needs, or they can be physical university campuses and work places, functioning side-by-side with virtual social networks such as Facebook, for meeting the human search for self-actualization, ranked at the highest level of the hierarchy of human needs. It is, therefore, in place to explore the question, in which type of space, physical or virtual, would persons be able to satisfy the needs which are at any of the specific five phases of needs? The examination of this question requires us to compare the two spaces from the perspective of human needs.

## Real and virtual spaces and human basic needs

As we have just noted, the human motivation for the fulfillment of basic human needs has always led to the construction of proper facilities in physical space, whether these have been fields, homes, working places or institutions (Figure 4.2). It has been for human activity over the centuries in real space, notably in science, engineering and education, to bring about the gradual emergence of virtual space during the last two centuries, and eventually turning it into a significant entity for human life. Electronic virtual space has gradually evolved since the invention of the telegraph back in 1837, currently maturing into the Internet as an open communications and information system, functioning as of 1994, and coupled later with the emergence of mobile broadband and the introduction and adoption of smartphones (see Kellerman 1993 and Chapter 1). The flexibility of the Internet, manifested in its ability to overcome space and time frictions and constraints, side-by-side with its becoming a second space for human operations, has reduced dramatically the well-known "tyranny of space/distance/proximity" (terms coined originally by Toffler 1980; Mitchell 1995; Blainey 1996 and Duranton 1999, respectively), which has been attributed to human action in real space.

As we noted previously, the Internet has constituted from its early outset a double media, in its constitution of both a communications system via emailing, coupled later on also with Web 2.0 for social networking, and an information system via websites installed on the Web. Thus, Internet-based communications,

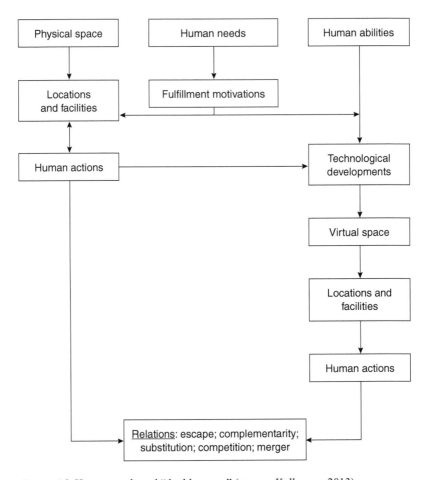

*Figure 4.2* Human needs and "double space" (source: Kellerman 2013).

written or audio-visual ones, transmitted through VoIP (Voice over Internet Protocol), may replace or extend face-to-face interactions which previously constituted the sole option for human instant interactions, being later on coupled with telephone calls (see Boden and Molotch 1994). The virtual sites, as their name implies, provide services and information that were provided in the past exclusively through specific real sites within physical space. Hence, the evolution of the Internet into an action space for the satisfaction of daily human needs, thus permitting the emergence of a variety of relations between physical and virtual spaces in this regard: complementarity; competition; substitution; merger and escape (see Kellerman 2006, 2012), which we will explore in the following sections.

Before dealing with these five possible relations between physical and virtual spaces let us return to Maslow's proposed hierarchy of human needs and take a look at it from a current spatial perspective, attempting to examine in which of the two spaces, physical and virtual, many people contemporarily satisfy the needs that are outlined for each level (Figure 4.3). Generally speaking, the more basic the needs, the more crucial is their provision in physical space and vice versa: the higher the ranking of the needs, the wider their possible gratification through the Internet. Thus, needs such as sleep, excretion and temperature are solely provided in physical space, but food, though eaten physically, can be bought also over the Internet, its preparation may be learned in virtual space and dietary information is available and exchanged through it.

Sex seems to be the most complex basic need as far as its satisfaction in virtual space is concerned. Sexual life over the Internet, the so-called cybersex, refers to sex that is generated and developed through the Internet, though the sexual activity per se is physical (see, e.g., Ben-Ze'ev 2004). Cybersex may be performed either through watching still and streaming pictures, or through written and video chats. Using the Internet for sexual activities constitutes one of its major uses, and abuses in this regard have turned into a major social concern. Thus, it was estimated for 2012 that some 12 percent of the websites are porno-graphic in nature and some 25 percent of the daily searches through search engine are for sexual materials and contacts (Zur Institute 2012).

Safety, too, in the sense of securing the human body and people's material property, is basically provided in physical space, but its management, as well as its control, may be based and operated through virtual space. In addition to the

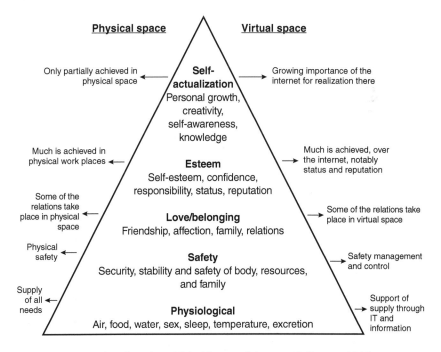

*Figure 4.3* Hierarchy of needs and "double space" (source: Kellerman 2013).

traditional need to protect material possessing, the contemporary virtual storage and flows of information over communications devices and media, notably personal and business information, require safety measures of their own, which are obviously provided virtually as well, and have turned into a major concern for Internet users at times of constantly growing cyber attacks. Thus, the popular Norton (Symantec) antivirus specialists reported for 2011 an annual growth of 81 percent in malware attacks since 2010, coupled with a 36 percent growth in Web-based attacks for the same year (Symantec 2012).

Climbing up the hierarchical ladder of human needs, when it comes to love and belonging, virtual space becomes more significant and even crucial as a contemporary channel for their provision for Internet users. Correspondence and talks, including video ones, are widely performed through the Internet and the mobile and fixed telephone systems, coupled with the near-disappearance of the traditional written communications through the postal system. The introduction of Web 2.0 in the 2000s has permitted the expansion of belonging opportunities even further through social networking platforms, such as Facebook, as well as through other tools such as mailing lists. This type of virtual belonging has turned into a daily routine for Internet users worldwide, and as we have seen already, over 40 percent of Internet users worldwide are currently Facebook subscribers (Checkfacebook 2013).

The higher level need of esteem has a lot to do with working places, which for most workers are still physical locations (Luca 2000, see also Chapter 7). However, even there, email correspondence, through both Internet and Intranet systems, notably within multi-locational companies, involve also matters of reputation and status that may be achieved, and even more so expressed, via the Internet. Growingly, and more frequently, virtual channels become the sole avenues for such expressions. Thus, within online communities, reputation may constitute a rather critical personal professional resource, as compared to more material resources (such as possessions) serving as resources for personal reputation in offline communities (Lampel and Bhalla 2007). Esteem may further constitute a need for social contacts, and virtual space has become a significant arena in this regard as well. Leading examples are social networks, such as Facebook and Twitter, which permit expressions of instant social acceptance, for instance through the counts of "likes" of posts sent by subscribers to their friends.

Finally, and at the highest level of the hierarchy of human needs, self-actualization may still continue to be, at least partially, achieved these days in physical space, but the Internet becomes crucial in this regard, notably for the obtaining of codified knowledge, mainly through websites, and tacit knowledge mostly through email correspondence. We will elaborate on this aspect in a following section. In addition, the Internet has become an arena for creativity and personal growth through the establishment of virtual businesses, such as via websites that sell information and virtual services (see, e.g., Capineri and Leinbach 2004). It is difficult, though, to estimate the extent of the virtual economy, given the differences in size between global giants such as Amazon, as compared to small single-person companies.

In the following sections we will briefly examine several possible relations which may have emerged between physical and virtual spaces, assuming that both spaces may permit the satisfaction of human needs in them: complementarity; competition; substitution; merger and escape. The identification of some of these trends, as far as the relationships between the two spaces are concerned, was initially offered from a mobility perspective, namely between moving in physical space as compared to moving in virtual one. Thus, Mokhtarian (2000: 1) distinguished among four possible relationships between telecommunications and travel: "substitution (elimination, replacement), generation (stimulation, complementarity), modification, and neutrality." Twelve years later, though, in 2012, it was still unclear which of these patterns dominates the relationships between physical and virtual mobility, though complementarity or generation seemed more acceptable (Aguiléra *et al.* 2012). Our concern here with the relationships between real and virtual *spaces* is different and possibly much wider than that related to physical and virtual *mobilities*, since our concern here focuses on the relationships between the two spaces with regard to the satisfaction of human basic needs, for which travel and telecommunications, or mobilties, are means and not objectives for action within the two spaces.

## Complementarity between real and virtual action spaces

Complementarity is the simplest possible relationship between physical and virtual spaces, at least as far as the provision of basic human needs is concerned. We noticed already briefly this kind of relationship regarding human physiological needs. All of the basic human physiological needs are initially met in physical space, but virtual space can complement physical space in the satisfaction of these needs in numerous ways and functions. It can, for example, serve as management frameworks and tools for food production and marketing. It can further serve as a major source for information on the characteristics of foods and their preparation, as well as a channel for food shopping, although customers may still prefer physical visits to stores and markets for food shopping (see Visser and Lanzendorf 2004). Regarding yet another physiological need, sex, we noticed already that virtual space has become a source for stimulation, imagination and sharing via websites, networking and email (see, e.g., Ben-Ze'ev 2004).

Complementarity applies not only to the most basic level of needs, the physiological one. It can further constitute an applicable relationship between real and virtual spaces for the satisfaction of needs at higher levels of Maslow's hierarchy. One can assume that for each of the higher level of needs the Internet has initially offered complementarity to physical space for the satisfaction of various needs. Thus, for example, belonging was channeled before the introduction of social networking through frequent face-to-face meetings complemented by postal letters and the more virtual telephone calls, and later on also through emailing (see, e.g., Bialski 2012). The introduction of social networking may have changed, in some way, the relationships between the two spaces from complementarity to competition with regard to belonging, since frequently virtual meetings may be preferred by interacting individuals over face-to-face ones.

## Competition between real and virtual action spaces

Competition between physical and virtual spaces, as far as the satisfaction of human needs is concerned, may emerge in several ways. One type of competition that could evolve between the two spaces we just noted with regard to friendship and relations, with the two spaces offering different modes for social interaction. Moreover, relations offered to individuals via Internet-based social networking may compete with their relations with other people taking place in the immediate physical space, though this does not necessarily have to be the general rule (Lee and Lee 2010). The instant availability of people over the Internet and the possibility to interact with them also through video may potentially, at least, compete with equivalent relations in physical space.

Competition between the two spaces may further emerge with regard to the reputation and status gained by individuals in their working place, between their daily face-to-face meetings with supervisors and colleagues, on the one hand, and their reputation and status emerging through Internet and Intranet networks pertaining to their work organization, on the other. The latter "virtual status" may permit, for example, the presentation of knowledge and skills by workers to their colleagues located elsewhere, which may be of a wider scope and significance as compared with similar presentations to their peers at their physical working place, in which their job status may imply the imposition of some limits as to the exposition and use of their skills (see Lee and Lee 2010).

## Substitution of real space by virtual one

The potential substitution of physical travel by virtual travel over the Internet, as well as the potential substitution among virtual media (e.g., between the telephone and the Internet) have been discussed elsewhere (Kellerman 2006a). The perspective here is wider, though, than travel and virtual mobility, relating to virtual space as a type of space, rather than merely a mobility platform, assessing its possible substitution for physical space. Graham (1998) proposed substitution as one of three potential processes for relationships between physical and virtual spaces (jointly with co-evolution and re-combination), based on the information technologies prevailing at the time of his writing.

Recent developments of technologies and applications for the use of virtual space may have brought about already, fully or mostly, the substitution of physical space by the virtual one, regarding several branches of the acquisition of codified knowledge. Such acquisition is one of the needs included in the highest level of Maslow's hierarchy, that of self-actualization. Thus, almost all scientific journals are now available through the Internet, and most of them have their old back issues digitized as well (Ware and Mabe 2009, see data in the next chapter).

Public data sources, such as census data, are also stored as digital data and are accessible through the Internet. Scientific books, notably older ones, are still mostly available only in physical libraries, although this may change in the future. The substitution of real space by virtual has emerged not only for the distribution of knowledge, but also for its acquisition through distance learning, or e-learning. We will discuss this option in Chapter 7.

## Merger between real and virtual spaces

Another possible relationship between real and virtual spaces for the satisfaction of basic needs is a kind of merger between the two spaces, in the sense that gratifying actions for some needs may take place simultaneously in the two spaces. A leading example in this regard is sexual needs, which can be motivated and fostered through video conversations or through Web pornography, bringing about sexual satisfaction of the body in real space. Another field of basic needs potentially leading to merger between the two spaces is status, when a message that is transmitted through email to a wide audience may yield appreciation and promotion in real space offices, at the physical location of the relevant person, simultaneously with emails coming in from remote locations.

## Escape from real space to a virtual one

The only geography branch that has explicitly related to Maslow's hierarchy is tourism (Pearce 1993; Hall and Page 2006). Moreover, tourism scholars claimed that "Maslow's hierarchy of needs is a key theory in travel motivation research" (Huang and Hsu 2009: 288), and that it "forms the basis for further development and applications to understand travel behaviour and demand for tourism" (Brown 2005: 481). Specifically, Hall and Page (2006: 47) ranked individuals' motivations for touristic activities, and following Crandall (1980), they suggested that leading the list of tourism motivations are the desire to escape from civilization and escape from routine work and responsibility. Of all the needs included in Maslow's hierarchy it was only self-actualization that was included in the list of touristic motivations and it was ranked 14th only, though it was noted that Maslow's needs should not be viewed as hierarchical but rather as simultaneous. Pearce (1993) further applied Maslow's motivations to groups of people, claiming that each group has different motivations and needs upon visiting theme parks for leisure. More generally, and of much importance for our discussion of virtual space, Iso-Ahola (1982) argued that tourism involves a double motivation: approach (seeking) and avoidance (escape), or in other words, pull effects of touristic destinations and push effects of home origins. Iso-Ahola (1982: 258) further argued "that tourism ... represents more of an escape-oriented than approach-oriented activity for most people under most conditions."

People's motivation for tourism has been explained, therefore, above all by a human need for escape from daily routines, which can be materialized on an occasional basis through moving physically from home and work environments to other, mostly but not necessarily, new, places/countries. However, Maslow did not consider this rather contemporary human need for temporary escape from routine activities and locations as one of the basic human needs.

The logic of physical escape via tourism may be applied to virtual space as well. Web 2.0 social networking tools, notably Facebook and Twitter, facilitate and permit daily, and for many of their subscribers most frequent, virtual escapes from their daily routines, as well as from the people located around them in physical space. This kind of virtual escape is available for networking people in their immediate physical environment, even on a continuous basis through smartphones, without any need for physical travel. Such frequent escapes from physical space to virtual ones constitute simultaneously a need and a one-way relationship between the two spaces, since opposite escapes, namely from virtual space to physical one, might be more seldom, if they exist at all. A similar, and on many occasions a rather individualistic experience of escape, excluding any socializing with other individuals, is offered by video games, which provide a "sensory experience" for players (Shaw and Warf 2009), employing technologies of touch for the reproduction of proximity and intimacy over distance (Paterson 2006). Thus, in 2010 some 36 percent of US Internet users engaged in online games (US Bureau of the Census 2012a).

Being continuously involved in escape activities through virtual networking does not necessarily nullify users' need for physical escape through tourism, since the latter constitutes an experience of a rather total nature for both mind and body: it is prolonged, and it involves simultaneous physical exposure to other places, people and events. Virtual escape has some special value for its subscribers. It might well be that the specific nature of social networking as a constantly available virtual escape for individuals while they still remain in the same physical location or move about it, explains the Facebook boom which, within a few years following its launching in 2004, attracted over 40 percent of global Internet users to subscribe (Checkfacebook 2013). Physical travel implies the use of the Internet, since preparations for it may involve search for touristic destinations through Web browsing, and this activity may be considered by actors also as some, more minor, escape from the routine physical environment.

## Exclusivity: novel human action over the Internet?

Under the contemporary circumstances of an existence of two spaces for a wide variety of human actions, one may wonder if it would be possible, some time in the future, that for the satisfaction of certain actions, virtual space will be the only possible action space. Therefore, there will be things for which there will not exist any option for performance at all in physical space. Would such a possibility imply new needs of human beings, which may only be met over the Internet? Such circumstances would amount to an exclusivity of virtual space as far as the satisfaction of such speculative future needs is concerned. Futurists may potentially respond to these questions by stating that it is too early as of yet to address them because the development of the Internet and virtual space at large has not yet reached saturation. However, it is of interest to briefly comment on these questions given the development of information technology so far.

The Internet permits to maintain conference calls, namely video conversations in which people from different parts of the globe, located in different time zones, interact. This experience cannot take place, by its very definition, in physical space. However, such an experience does not necessarily call for any changing human needs or for the emergence of new human emotions, differently than those of people involved in a conference conversation while sitting in separate rooms in the same building (Boden and Molotch 1994). More generally, then, the Internet as well as sophisticated telephone systems permits to overcome basic time and distance barriers, but this does not imply a change in basic human needs. The overcoming of space and time constraints have become a routine for many workers, thus bringing about more intense working space and time, notably when coupled with mobile broadband availability through smartphones. However these contemporary trends do not seem to point to any exclusivity of virtual space with regard to the required action space for any activities, or for the satisfaction of any new human needs.

## Space as a human need

So far we have related to space, in its two categories of real and virtual spaces, as a platform and thus as a basic means for the satisfaction of basic human needs. We may, though, look at real space also from the angle of its being a basic need by itself for human beings (as well as for any other forms of life). From a macro societal perspective real space may be viewed in this regard as a container for all humans, as well as its being a container for their activities and products (Kellerman 1989). Extending this approach to a rather micro perspective for individual humans, for each individual real space would be viewed as serving as a physical anchor, which facilitates their very material existence on earth. This anchoring of humans by space may lead to personal feelings of territoriality as well as to the creation and/or conception of personal spaces (see, e.g., Gold 1980), and it may further permit and facilitate the emergence of human action space. However, real space in its most basic constitution as a need for humans constitutes its being an anchor for the very existence of the live human body. Space, as such, has not been included in the list of basic needs outlined by Maslow (1943, 1954, 1968), maybe because it was taken for granted, but in its nature as a most basic anchor for the human body it should have been classified jointly with the most basic physiological needs.

Turning now to virtual space, what about it as possibly constituting a need as an anchor? Once virtual space has turned into an action space, notably through the Internet, has it become also a need by itself, like real space, or is it merely a platform facilitating the satisfaction of human needs, because humans continue to be anchored in real space? One way to look at this question would be to imagine a hypothetical total collapse of the Internet bringing us all back to the pre-1994 state of affairs. From a wide societal perspective, all the classes of human needs could then be satiated through traditional space, in slower, less efficient and less sophisticated ways, as things were done until the evolution and adoption of the Internet, and thus nullifying the possibility of looking at virtual space also as a need. However, from the individual perspective of a growing number of Internet users, notably but not only young ones, things are a bit different, as many users have become addicted to Internet use, mainly for social networking, thus regarding it a virtual, but still basic, anchor for their life. Assuming, then, that virtual space growingly becomes a need by itself, at least for some, where should it be categorized in the hierarchy of basic needs? Should the most basic level of physiological needs, be expanded, so that it will include also basic psychological needs? These are questions which may require further attention if current trends of Internet use intensify.

## Conclusion

The basic human needs of physiology; safety; love/belonging; esteem and self-actualization have recently come to share their fulfillment between physical and virtual spaces, brought about by the contemporary maturing of Internet and communications technologies, thus permitting human action in double space. The role of virtual space in this regard grows along the hierarchy of these needs, moving from secondary importance for the most material needs, the physiological ones, to a major importance in the most abstract ones of knowledge acquisition for self-actualization.

This growing role of virtual space for the satisfaction of human needs is expressed in hierarchical relationships between the two spaces along the hierarchy of human needs. Thus, in the lower levels of the hierarchy, virtual space offers mainly complementarity to physical space, with the latter constituting the crucial arena for the satisfaction of the most basic needs, moving to competition with physical space in the higher levels of the hierarchy, and eventually substituting physical space at the highest level of the hierarchy. These rather diversified and simultaneous relationships between the two spaces regarding the fulfillment of human needs are significantly different than the singular relationship of complementarity proposed at the time for the relationships between physical and virtual mobilities.

A special relationship of escape from physical to virtual space has been brought about by social networking, which is similar to the one offered by tourism through temporary physical travel from place to place. This virtual escapism through networking is new, though being widely and swiftly adopted, and we will address it in detail in Chapter 8. It is too early, though, to speculate on future routines that may emerge and be adopted by individuals with regard to such virtual escapes. Further with regard to the future, it does not seem real to foresee that virtual space will offer any exclusivity for the fulfillment of as yet new and unforeseen human needs. However, the Internet increasingly becomes a human need in itself, beyond its constitution of a platform for need satisfaction, similarly to real space serving as an anchor for the very existence of humans.

# 5    Curiosity and its satiation

In the previous chapter we discussed the role of the Internet for the satisfaction of human basic needs, as defined by Maslow's hierarchical list. In this chapter, the Internet will be presented as a new and most powerful means for the satisfaction of yet another human basic need or nature, namely that of human curiosity. In some ways, this chapter is, therefore, a direct continuation of the previous one, focusing here on a basic human need or nature, curiosity, which was not included in Maslow's hierarchical lists of such needs. The acquisition of knowledge, ranked high on Maslow's list referred to the need for knowledge in order to achieve self-actualization, may bring about status and power, but he did not refer to knowledge as something which is sought for because of one's curiosity. At the lowest level of his hierarchy, Maslow referred to food as a most basic human physiological need, but not to information and knowledge as such, though as we will see, a hunger for information might be related to hunger for food.

In the following sections we will explore the biological and psychological roots for curiosity, and its classification into prudential and epistemic curiosities. Physical space will be exposed as a trigger for curiosity in many ways, such as landscapes, specific designs of classrooms, side-by-side with websites and social networking. Furthermore, real space will be presented as a curiosity objective by itself, inviting its visitation by curious explorers. However, the most significant role of real space for curiosity is its constitution as means for the satiation of curiosities of all kinds. Traditionally, intellectual curiosity has been satiated in real space through library visits in the search of codified knowledge, as well as through face-to-face contacts with colleagues looking for tacit one. We will see in this chapter that the Internet has changed the direction of information and knowledge flows: when using the Internet for knowledge acquisition knowledge flow towards curious people rather than the traditional opposite way of people physically seeking information through their library visits, etc. The Internet serves, therefore, as an action space for intellectual activity, and as an action space for the satiation of curiosity. In addition, the Internet, as email platform, permits wider collaborations among scientists in their development of new knowledge through research. Thus, we will see that in the information age, the intensity of scientific or epistemic curiosity, measured by the number of articles per scientist, has not grown, but the rate of collaboration among researchers has increased significantly.

Curiosity and its satiation constitute basic and constantly present processes in the daily lives of all human beings. By the very nature of their work, cycles of curiosity and its satiation are even more striking in the activities of scholars whose work implies foremost the asking of questions and the posing of problems, followed by attempts to explore and find possible answers and solutions for them. Most of us tend, therefore, to take curiosity and its satiation for granted, and when becoming curious about the understanding of curiosity and its satiation, as human needs and occurrences, we would turn mainly to psychology and biology as the pertinent disciplines for coping with these aspects of human life.

We will attempt in this chapter to trace the spatial dimensions of curiosity and its gratification, which will lead us to the special role of the Internet in this regard. Psychology and biology literatures focusing on curiosity have tended to ignore the spatial dimensions of curiosity, since curiosity, by its very nature, seems to constitute, at a first glance at least, a non-spatial human need. Still, however, curiosity was claimed to have a geographical nature: "When people try to make curiosity explicit, they tend to speak and write geographically. Curiosity is widely portrayed as a desire to encounter the unknown, articulated with reference to various *terrae incognitae*. This is expressed through travel writing and exploration narratives" (Phillips 2010). As we will see, despite this general claim for the role of geography in curiosity, the attention of geographers to curiosity and its varied spatial manifestations has been rather scant. Thus, our following discussions will be rather exploratory in nature, attempting to outline some basic terms, as well as several contemporary processes, based on some general data, which will permit us to highlight the role of the Internet in curiosity and its satiation.

Before delving into the discussion of space and curiosity we will devote some attention to curiosity in general, focusing on its nature, causes and drives, and this will be followed by the highlighting of the spatial dimension of curiosity from three angles. First, the possibility that specific places and landscapes serve as curiosity triggers for individuals located in them, in the sense that curiosity is made possible through these places or landscapes as providers of some stimulating contexts and environments, thus making people become curious for things which are not necessarily related to or included in the spaces which may invoke these curiosities. Second, the option that new and yet unexplored places by people serve as objectives for their curiosity and its satiation, eventually making them go, see and experience these places. Third, and most widely relevant, once curiosity emerges regarding almost any non-spatial or spatial aspect of life, space may serve as means for the movements of people to sources of information and knowledge in their seeking of gratification for their curiosity.

Summarizing these three spatial dimensions of curiosity, space may possibly serve as trigger for curiosity but space is not a precondition for its arousal. Furthermore, space may constitute an objective for curiosity, but curiosity per se is much wider than the urge for exploration of new places, since people may get curious about endless non-spatial things. On the other hand, however, the satiation process for all types of curiosity almost always involves physical or virtual movements of curious persons in their search for knowledge or information for the satiation of their curiosities.

Following the exposure of the three angles of space for curiosity, as trigger, as objective, and as means for its satiation, we will attempt to combine them all into a single general framework, outlining the spatial dimension of curiosity. Finally, these discussions will be followed by a concluding discussion on the ramifications of the contemporary virtual means for the satiation of curiosity.

## The nature of curiosity

The Oxford Dictionary linguistically defines the word curiosity as "a strong desire to know or learn something" (Pearsall 2002). In English, as well as in most European languages, the term curiosity stems from the Latin *curiositas*, but in German there are two words for curiosity, with a differentiation, which is important for us. The first word is *Neugierigkeit*, referring to the rather general human eagerness to know something, a kind of curiosity which Gade (2011: 10) called *prudential curiosity*. The second German word for curiosity is *Wissbegierde*, which Gade (2011: 11) termed *epistemic curiosity*, and it refers to intellectual curiosity, or to thirst for knowledge (see Gade 2011: 4, 20). People who are professionally involved in epistemic curiosity were termed by Gade (2011: 10) as *professional questors*, and by Pasternak (2007: 128) as *homo quaerens*. This group of professional "questors" consists mainly of scientists who are professionally engaged in epistemic curiosity and its satiation, but also of others, such as journalists and investigators. Science itself was, therefore, called the *discipline of curiosity* (Groen *et al.* 1990).

From a scientific perspective, curiosity was defined as "the desire for information in the absence of any expected extrinsic benefit" (Loewenstein 2002: 1), namely a "pure" desire for information leading to exploratory behavior (see Fowler 1965). In daily life the term curiosity is used, though, in a wider and more permissive sense, relating also to general human urges for information, and which can be tied to some specific benefit. For scientists, as well as for people involved in learning at large, the desire for knowledge constitutes a rather general framework within which there nest and evolve more specific cycles of curiosity and its satiation. By using the term *cycle* here we refer to the posing of research questions or hypotheses and the proposing of answers to them. It was attributed in this regard to the Irish novelist Laurence Sterne, that: "the desire of knowledge, like the thirst for riches, increases ever with the acquisition of it" (e.g., goodquotes.com 2013).

Aristotle made an early reference to curiosity, opening his book *Metaphysics* by stating that "all men by nature desire to know" (Book I Part 1: 1). This curiosity implies further that "humans seem to have an inborn tendency to be explorers" (Hägerstrand 1992: 35). More specifically, curiosity aims to explore a space that must still be furnished for us. With questions and gestures more spontaneous than goal-oriented, curiosity explores what it does not yet know and what seems interesting and worth knowing, often for reasons it cannot name.

(Nowotny 2008: 3)

Furthermore, it is not only that we do not always know why we get curious about something, but sometimes we do not search for the sake of some specific objective: "We search not only out of necessity, but out of sheer curiosity as well" (Pasternak 2007: 117).

Numerous origins for human curiosity have been proposed by a number of scholars. Biologically, the natural tendency of the brain to perform as a cognizing organism was suggested as constituting a primary source for curiosity, and curiosity may, thus, be considered as a basic human need like eating (James 1890, see also Gade 2011: 9). From both physiological and psychological perspectives, curiosity may further constitute a secondary drive, stemming from more basic ones such as hunger. In addition, curiosity may be considered a primary drive for resolving certain uncertainties, in ways different than the mind's resolution of other uncertainties. For example, in the case of hunger, satisfying it through the eating of one food makes all other foods unappealing following bodily satiation, but satiating intellectual curiosity regarding one specific issue does not reduce curiosity for other issues (see Kellerman 2006, 2012a). Yet another, more psychological, explanation sees curiosity as arising from a gap between existing and desired knowledge (see Fowler 1965 and Loewenstein 2002 for reviews). Despite all these human needs for curiosity per se, Gade (2011: 9) claimed that "those who have an intense desire to seek out, explore and understand new things in their environment constitute a small minority of any population," and this might relate mainly to epistemic curiosity.

One cannot normally reach a condition of full rest from curiosity, only times of relative rest from curiosity, whether one is awake or at sleep. Such times of low curiosity may amount to a form of *apathy* in the sense of incuriosity (see Fowler 1965). Apathy, with the connotation of incuriosity as a required human condition for some rest from curiosity (rather than apathy as a personal characteristic), is, thus, a relative condition. The deepest condition of rest/apathy is achieved during sleep, but sleep may produce unconscious dreams, which may reflect curiosity. Also, one may rest from one type of curiosity, e.g., resting at home from work-related curiosity, only for becoming curious, then, for some personal, domestic, or any other topics that are not work related.

## Space and place as curiosity triggers

The specific locations of individuals and their environmental contexts at any given time and location may bring about some curiosity. Such places might be as banal as daily visits of a railway or bus station with riders discovering there a new sign or advertisement raising curiosity for its meaning, or seeing in the station a person dressed in some way bringing others to wonder about the qualities of this clothing. Physical mobility or one's moving to specific locations, notably through walking, may further bring about curiosity evoked by a wealth of information exposed to while walking, information which is not necessarily sought for by the moving person, no matter whether the travel objective is for the reaching of people, places, or events. This information consists mainly of the rich spatial contextuality (e.g., buildings, people, stores, etc.) incorporated into the physical paths used by individuals when involved in corporeal movements.

Space and its perception as curiosity generator or trigger is specifically manifest in spatial units devoted for learning, such as real classrooms or virtual e-learning websites, the design of which may potentially generate curiosity among students (Phillips 2010). Thus, "geographers have a distinctive contribution to make by illuminating the ways in which environments variously encourage and discourage curiosity" (Phillips 2010: 448), a contribution which may be shared also by environmental psychologists and architects.

Virtual space, or the Internet, may serve as a curiosity trigger, even beyond the design of e-learning websites, through the facilitation of communications among people who share areas of interest, as Gade (2011: 144) noted: "Internet technology has fostered a bandwagon mentality by the creation of networks set up to share research interests." As we will see in a later section, the Internet constitutes, foremost, a huge source of information and knowledge for curiosity gratification. The very availability of almost any possible information sources over the Web has facilitated untargeted curiosities for information and social reaching-out through browsing and networking, respectively. To paraphrase McLuhan's (1964) famous phrase on media at large, the Internet is not just at once the medium and the message, as it constitutes a third dimension as well in its universal availability as an information source, thus turning it into a motive for curiosity, side-by-side with its being a channel for exploratory behavior in the search for answers.

The cyberspatial Internet has further served as a platform for the most extensive emergence of global and borderless social networking among people (see Kellerman 2012 and Chapter 8). Thus, as we noted already, some 43 percent of the global Internet users were also Facebook subscribers in 2013, amounting to 12 percent of the global population! About one half of these subscribers logged on to Facebook daily (Facebook 2012; Checkfacebook 2013; see also Kellerman 2007). Obviously, and by the very nature of networking, this system of new and diversified networks provides an enormous source for interactive curiosity cycles of questions and answers among interacting persons.

## Space as curiosity objective

Space has long served as an object for exploration for humanity at large, as well as for individuals curious about visiting places and landscapes which they have not yet visited. Thus, for John K. Wright (1947: 1) it was "*terra incognita*: [these] words stir the imagination." Geographers, more than other scientists, possess "an imagination peculiarly responsive to the stimulus of *terra incognita* both in the literal sense and more especially in the figurative sense of all that lies hidden beyond the frontiers of geographical knowledge," and "curiosity is a product of the imagination" (Wright 1947: 4). Thus, according to Wright, unknown spaces stir the imagination of geographers for wide as well as deep explorations, simultaneously amounting to curiosity and yielding one, as well. Over the generations, "men's quest has encompassed the entire surface of the earth; he has been to the moon and observed the most distant stars" (Pasternak 2007: 114). Not only do people attempt to discover new spaces, but, as Phillips (2010) stated, the urge to discover and explore literal space has led to the metaphorical use of geographical language in this regard for explorations of unknown non-literal spaces as well: "when people try to make curiosity explicit, they tend to speak and write geographically" (p. 449), using, for example, the term *terra incognita* for non-spatial unexplored themes (see also, e.g., Nowotny 2008: 3). Still, though, it was believed that geographical curiosity is "more narrowly focused than mankind's; it is also more conscious, orderly, objective, consistent, universal, and theoretical" (Lowenthal 1961: 242), and in this reiterating Sauer's (1941: 4) claim for geography to constitute "focused curiosity" (see Jackson 1980).

Attempts to discover new spaces and places are not restricted to geographers only, who do so professionally. Going to new places is something which most people would like to do in varying frequencies. The ability to travel by car and plane domestically and internationally has made room for exploratory travel through vacationing. Some of the overall contemporary growth of tourism may have to do with this kind of curiosity.

Both geographers and individuals who are engaged in the planning of touristic visits tend to make use of the Internet, albeit as an auxiliary service. Geographers may make use of maps posted on the Web, as well as of textual websites related to their studies. Tourists may consult promotional and advisory websites related to their planned destinations, and they growingly place reservations for travel and lodging through the Internet (see Chapter 7 for discussion on travel online).

## Spatial means for the satiation of curiosity

So far we have related to the spatial dimensions of curiosity per se, discussing the role of space as curiosity trigger and/or as curiosity objective. The discussion in this section turns to the role of space in the satiation of curiosity of all kinds. This role of space as means for curiosity satiation is crucial for the human attitude to curiosity per se, since being able to satiate curiosity may avoid frustrations in dealing and coping with curiosity. Curiosity has become of special importance in contemporary life mainly because of our new abilities to satiate curiosities instantly and virtually through the wealth of information and knowledge swiftly available over the Internet, thus expanding tremendously the avenues and abilities for our curiosity satiation. It is, thus, the growing ability to satiate curiosity that may permit people to be more curious and wonder about more diversified issues, and it is, therefore, for the Internet, and notably the Web, to be considered as an action space for the satiation of curiosities.

As we noted already, prudential curiosity constitutes a human drive that relates to the daily needs for information, covering human life in all of its aspects. The Web is, by its very nature, an endless action space for the satiation of such daily curiosities for anything ranging from news to flight schedules, bank account balances and astrological forecasts. However, for certain types of desired information one might achieve full satisfaction only when physical face-to-face contacts are established, even if the travel effort involved in the reaching of such gratification may turn out to be immense. Of special importance in this regard, is people's tendency to invest extensive travel efforts when close social/family relationships are the case, or when business affairs call for them. On the other hand, however, acquiring information on the balance of a bank account is the same whether it is received through the Internet or through walking into a bank branch.

When it comes to epistemic/scientific curiosity, "more and more means and instruments, mostly but not entirely scientific and technical in nature, are at our disposal to expand the space of our experience" (Nowotny 2008: 3). Over the generations, traditional academic libraries have served as the major means for the satiation of scientific curiosities for information and knowledge, becoming challenged contemporarily by the Web which provides unprecedented options for the satiation of epistemic curiosity, through its instant and universally available, updated, and extremely varied and multi-sensory wealth of information.

At the time, academic libraries constituted the founding facilitators of free access to knowledge, in their serving as active depositories for journals and books. As such, they have been centrally located within campuses in remarkably designed structures, in order for them to be physically striking structures within university campuses. However, in the Internet era libraries are in the process of growingly becoming archives for old journals and books, and they might potentially turn into consulting and discussion centers for the new art of information and knowledge retrieval, mainly of virtual resources. Specialized library collections, such as those devoted to ancient books and manuscripts, are also in the process of being scanned and thus accessed electronically. Hence, as Daniel Greenstein, Vice-Provost of the University of California, predicted: "the university library of the future will be sparsely staffed, highly decentralized, and have a physical plant consisting of little more than special collections and study areas" (Kolowich 2009).

The location of scholarly work, or the work places of scholars, as far as the need to satiate scientific curiosity is concerned, may become much more flexible through the use of the Internet, since information and knowledge resources flow now to scholars through the Internet, rather than vice versa, namely scholars going to libraries, as was common traditionally. This is notably so, since the publishing industry has taken advantage of information technology and the universal availability of the Internet to scholars, by turning journal publication almost completely electronic. Unfortunately, the high prices of library subscriptions to journal databases have brought about growing gaps between universities in the developed and developing worlds, as far as access to the journal literature is concerned. Among scholars, notably geographers in the developed world, this issue of pricing of electronic publishing has been considered a political issue (see, e.g., Pickerill 2008). Back in 2008 some 96 percent of STM (science, technical and medicine) journals and 87 percent of arts, humanities and social sciences journals were online (Ware and Wabe 2009). Interestingly enough, the popular magazine industry, which could have made use of the same technologies as those of the scientific journal industry, still publishes magazine issues mostly in print only. This difference in preferred publishing modes may stem from a difference in journal reading habits between scientists and laymen. Whereas scientists normally read specific articles as separate reading units rather than reading full journal issues, readers of popular magazines may tend to read or browse through full journal issues as single units, which consist internally of numerous articles mixed with advertisements.

More recently, e-journals have been followed by the publication of e-books in general, and more partially so also of scientific e-books, normally side-by-side with the publication of printed editions. Here too there is a pricing issue, since the prices of scientific e-books are frequently as high as those of printed books, accompanied with a highly restricted permission for page printing, thus permitting e-book purchasers to print only a small number of pages. In the e-book industry the trends among laymen and scientists are reverse as compared to those for e-journals. Thus, the sales of the general e-book industry in the US surpassed those of printed books already in 2010 (YUDU 2011), whereas for academic publishers the sales of e-books amounted by then to merely 8–10 percent of their total income from book sales (Neilan 2010). The general e-book market was boosted with the development of specialized computers for book reading, led by Kindle (Amazon); Story HD (Google); iPad (Apple); and Nook (Barnes and Noble). However, it seems that printed books still offer more flexibility to academic readers, but this may still change along with technology developments and industry adjustments.

Another important development with regard to the location-free availability of literature has been the digitization of older books making them available freely and globally over the Web. Google has initiated the Google Books project for the scanning of old books for which copyrights have expired, and it did so in cooperation with leading academic libraries in the US and UK. The courts stopped the full digitization of more recently published books, however (Darnton 2011). There are, however, national digital libraries in the making in several European countries (e.g., Norway and Netherlands), as well as for Europe at large (Europeana). In addition, there have emerged book digitization projects which are not country specific but rather language and/or culture specific, such as the Bar-Ilan University Responsa project (Israel) and the HebrewBook (Canada) project, digitizing Judaic sources and writings authored over the last two to three millennia.

These seemingly technological changes that have revolutionized the means for the satiation of curiosity in general and of epistemic one in particular may involve also some change in cultural values. Printed books have always constituted a material symbol for knowledge and culture, and religious scriptures may have frequently also possessed a touch of sanctity. It is still questionable whether e-books will change these values. Wieseltier (1998: 155) noted on information/knowledge relations that: "I have no doubt that this technology [CD-ROM] will debase knowledge and reduce ... all the materials it holds, to the status of information." It is true, on the one hand, that contemporary information technology permits easily the search and finding of specific pieces of knowledge/information within a book through search engines, thus accentuating the role of browsing at the expanse of that of reading, and thus avoiding the full study of whole books. However, on the other hand, by collecting numerous widely spread pieces of information in the literature through search engines, researchers may be widely aided in their efforts to create new knowledge, something which might have been difficult or next to impossible to produce in the past when book indexes were the only search toll available on a book-by-book basis.

## Trends in epistemic curiosity and its satiation

Knowledge inputs by scholars can measure the transitions in the behavior of scientists when attempting to satiate their epistemic curiosity at times of wide availability of the Internet as action space for researchers, namely the reading pace of articles, as well as by knowledge outputs, namely potential changes in the number of scholarly publications per scientist. An additional output dimension is the rate of collaborative studies and resulting publications permitted by the communications mode of virtual action space, namely the use of emails, video conversations and other electronic communications media for communications among cooperating scholars.

The reading pace of papers by scientists has increased in recent years, mainly due to the incorporation of search engines and other information technology tools into electronically accessed articles, thus turning electronic articles from pieces of knowledge into manipulated knowledge/information devices. Hence, the number of articles read annually by scientists rose from an average of 150 in 1977 to some 270 in 2004–2006, though the average length of articles increased from 7.4 pages to 12.4 during the same time period. Coupled with the growing number of papers read by scientists and their growing length, the time spent on article reading decreased from 45–50 minutes per article to merely 30 minutes, attesting to the possible use of search engines for browsing through some of them rather than reading them all (Tenopir *et al.* 2007; Renear and Palmer 2009; Ware and Mabe 2009).

Interestingly enough, however, the recently increasing inputs by scientists, as measured by the number of articles they read in the process of scholarly curiosity satiation, have not necessarily yielded increased outputs, when measured by a simple counting of the number of publications per scholar. It turns out that over the last two centuries the number of scientific journals has increased steadily at an annual rate of 3 percent, side-by-side with an average annual growth rate of 3.5 percent in the number of articles. These growth rates have been coupled with an annual average growth of 3 percent in the number of publishing scientists! (Mabe 2003; Ware and Mabe 2009). When the Internet was still new, notably the Web, those scientists who made use of it tended to produce more articles (Kaminer and Braunstein 1998). However, this early advantage has been lost soon after, as of the late 1990s, when the Internet became universally used by the academic community. The Web has served scholars in yet another way, namely for the interim and final presentations of research projects, side-by-side with their publication in journal articles, and such Web publications were made through websites of the research projects, as well as through dedicated pages on websites of universities and research centers (Fuller and Askins 2010).

The major change in scholarly article authorship as a result of the vast adoption of the Internet as a scholarly action space, aided by information technology tools, has rather been growing research collaboration among scholars, both domestically and internationally. Thus, the share of co-authored articles worldwide increased from 40 percent to 61 percent between 1988 and 2005, and the number of authors per article increased correspondingly from 3.1 in 1988 to 4.5 in 2005 (NSF 2008; Ware and Mabe 2009). This growing collaboration among scientists has resulted in a significant decline in the average annual productivity per unique author, going down from 1.0 paper in 1954 to merely 0.7 in 2000 (Mabe and Amin 2002; Mabe and Ware 2009). These findings and trends of growing collaboration among scientists in article authorship while keeping the general number of articles steady have been reported extensively (see, e.g., Ding *et al.* 2010 for a review). Still, Abt (2007: 286) noted in this regard:

> I do not know why the average annual number of papers per scientist has remained constant during two generations of scientists, particularly with the preparation of manuscripts being made much easier through the change from typewriters to computers. Of course with greater content per paper now, the more theoretical discussion and modeling and more data, all take more work (and perhaps more authors) to produce a paper than in the past.

Unfortunately, it is almost impossible to assess the quantitative trends of steadiness in scientific publication productivity side-by-side with the growing collaboration among scientists through a framework that would evaluate potential trends in the quality of scientific articles.

The trend of steadiness in scholarly publication productivity seems to further reflect a lack of increase in the number of cycles of epistemic curiosities and their satiation by scientists at the time of massive adoption of the Internet jointly with the adoption of other information technologies. Rather, they have yielded more complex cycles of curiosity and its satiation, through increased collaboration among scholars in the information age. Such growing collaboration implies more sophistication in scientific work procedures, again facilitated by information technologies. Collaborations among scholars require also face-to-face meetings in addition to ongoing email and video contacts. Though no data on faculty travel could be found, its crucial importance has been demonstrated (Staats *et al.* 2006).

The introduction and adoption of information technologies and media have brought about geography of research that has become more complex in the sense that, on the one hand, it has become easier to reach information and knowledge through the Web from anywhere globally. On the other hand, it has become both more attractive and more demanding to produce new knowledge through collaboration among scientists from various universities, countries and sometimes even disciplines, using emailing for joint research projects. Thus, the two components of the Internet, the Web as an information/knowledge depository, and the email system, as communications means, have both contributed to some change in the geography of scientific work or in curiosity/satiation cycles. These have amounted to a revolutionized action space for scientists with their physical location becoming of lower importance, as long as wide access to the Internet and to journal collections accessed through it are available. This new geography of increased mobilities of researchers is in line with the more general societal trends in the information society (see Kellerman 2012a).

## Wider assessment of contemporary satiation of epistemic curiosity

One may view epistemic curiosity and its satiation from a more general perspective of processes of stimulation leading to production. As we noted already, when the Internet has turned into a major mediating means between the two elements of curiosity (as stimulation) and its satiation (as production), then productivity (measured by the number of articles) has not grown but production patterns have become more complex socially (through joint articles) and geographically (through growing locational flexibility for work). It is intriguing to take a look at economic production in general from such a perspective, so that capital gains and profits will serve as the stimulation, equivalent to curiosity, and products and services will serve as outputs, similar to curiosity satiation. One may thus wonder if the introduction of information technology at large and the Internet in particular which have facilitated also an extended globalization of production, have increased productivity in economic production. As Gordon (2010) and Cowen (2011) have shown, this is not the case, neither for the American economy, which was the first to widely adopt the Internet, nor for the global economy at large. Gordon (2010) further claimed that the economic impact of the Internet, in terms of the possibility of bringing about higher levels of productivity rates, will be felt only several decades from now, a trend which bears similarity to the model of delayed productivity growth proposed at the time by David (1990) for leading inventions, such as the dynamo.

Interestingly enough, whereas the direct contribution of the Internet to productivity growth might still be modest, its indirect contribution may turn out to be of higher significance and in a paradoxical way. It was shown that the use of the Internet by workers for private purposes while being at work, the so-called *workplace Internet leisure browsing* (WILB) (Coker 2011), or as termed by others as *personal web use* (PWU), as *cyberloafing* or as *cyberslacking* (Polzer-Debruyne 2008), may increase productivity! Coker (2011) has found that WILB of up to 12 percent of work time may yield an increase of up to 9 percent in productivity scores.

Our observations on trends in the information age regarding scientific productivity stimulated by curiosity and the simultaneous and still developing and complex trends in the wider world of economic production stimulated by economic drives, seem to hint to processes of change which are still unsaturated, as the new technologies open up new horizons and ways of work.

## A general framework for the spatial dimensions of curiosity

Following our discussion so far we can present the role of space for curiosity through a simple model (Figure 5.1). Curiosity in general, including both prudential and epistemic curiosities, may arise, among other stimulants, through space as a trigger, whether the spatial unit is a routinely used one such as a classroom, or whether it is a new place for somebody, the exposure to which may bring about some curiosities. Space can further serve as an objective for curiosity, when some eagerness to learn of or visit new places emerges, either with laymen or professionals. Physical visits to places engage now not only transportation technologies, but also may further involve the use of navigation technologies, mainly GPS.

The quest for curiosity satiation has always had a spatial dimension, since the satiation process for prudential curiosity may imply physical visits to some facilities, for instance going to a store and trying on a dress there, following exposure to advertisement on TV. For epistemic curiosity, talking face-to-face with colleagues may present the exchange of tacit knowledge or information for the satiation of some specific scientific curiosity, coupled with visits to libraries for reaching codified knowledge through books and journal articles. In more recent times virtual visits for finding tacit information/knowledge have emerged through virtual communications means such as phone calls and emails, side-by-side with Web surfing for accessing codified knowledge. Both laymen and scholars perform all these searches for information and knowledge. As we noted already, libraries, as depositories of books and journals for the satiation of curiosities, are weakening in their role, since published materials have become available electronically, thus providing location-free availabilities to readers. By its very nature, scientific curiosity is normally only partially satiated through existing codified materials, since these materials serve as benchmarks for the production of new knowledge. The invention and adoption of the Internet have brought about growing collaborations among scholars. Needless to say that the very nature of scientific work implies that the satiation of some epistemic curiosity, phrased into research questions or hypotheses, or the completion of a curiosity cycle, may bring about a new one.

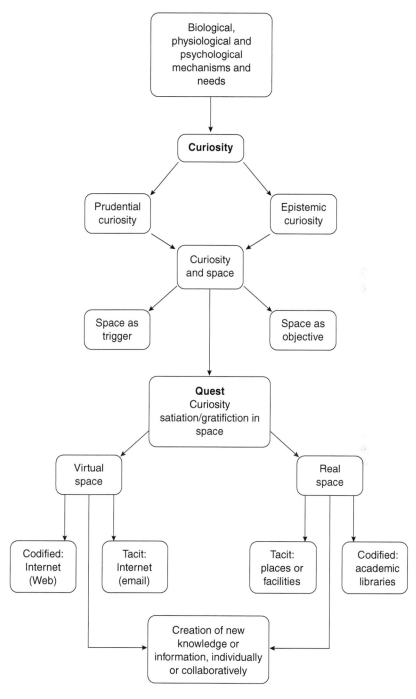

*Figure 5.1* A general framework for the spatial dimensions of curiosity.

## Conclusion

All the three spatial dimensions of curiosity (trigger, objective and means) have undergone some major changes in the information age, dominated by the emergence and wide availability of the Internet, at least in developed countries. Thus, virtual websites and their attractive design, side-by-side with growing networking, have provided extensive triggers for curiosity. The growing possibilities to move physically and tour other places and countries have made it possible for people to be more curious regarding other places and have made it possible for them to satiate such curiosities by physically visiting them.

It seems, though, that the major change regarding curiosity in the information age relates to the availability of novel means for its satiation, mainly the Web and its enormous knowledge and information resources. These resources have become available to scholars, the professional "questors," as well as to laymen. The digitization of both tacit and codified knowledge, through emailing and the Web, respectively, has brought about more complex research patterns through growing collaborations among scientists, rather than potentially bringing about an increase in their productivity, measured by the number of their publications. The dependence of scholars on virtual mobility as compared to their past dependence on their physical locations in campuses may attest to the recent statement that "it seems that [for them] the tyranny of space is being replaced by the *tyranny of mobility*" (Ferreira *et al.* 2012: 688). The trends of growing locational flexibilities in the work of scholars as permitted by the Internet are similar to trends in economic production at large, which has become more and more dependent on email for connectivity and on the use of websites for marketing.

Generally then, it seems that the nature of scholarly work, or the satisfaction of scholarly epistemic curiosity, has undergone dramatic transitions in the information age so far, in turning the Internet into a second action space, sometimes even the primary one. This has emerged for all of the major phases of research: the search for relevant literature, mainly through digitized articles, through the Web, and partially also through digitized books; the performance of research, notably cooperative one, with emailing serving as the major communications platform; and the publication of research results, once again in digitized journals. Furthermore, the digitization of research resources has facilitated locational flexibility of scientists, as compared with their previous fixed locations, in the close vicinity of academic libraries. Laboratories are still located in real space, but experiments and their results are usually available in digital formats, and thus are also being transmitted over the Internet.

An open question is whether non-professional "questors" in the Internet age, involved widely in prudential curiosities in their daily lives, may experience an increase in the number of curiosity cycles, as a result of the availability of the Web for instant and extensive satiation of their prudential curiosities. A related question is whether laymen have extended the topical range of their prudential curiosities, again given the availability and accessibility of the Internet for their satiation.

# 6   Personal identity

What revolution am I talking about? It is the revolution taking place with the digitization of identity, the wedding of selfhood and the electronic age.

(Mills 2002: 69)

Personal identity constitutes simultaneously a personal internal need and an external expression. The external expression of personal identity may, frequently, reflect one's internal personal identity, and it may come into external expression through one's presented identities in social networking over the Internet. However, at frequent other times, people's expressed personal identity involves some basic information about them, as well as their credentials, becoming available to others through institutional and personal homepages, as well as through other third-party information sources, such as websites specializing in the supply of such information. These availabilities of personal identity to others may be of importance to the individuals who expose their identities mainly for work as well as for social networking purposes. For those who seek information on the personal identity of others, these availabilities may satiate their curiosity for such personal information.

The two forms of personal identity, constituting of internal and credential identities, which are both presented and available over the Internet, will be highlighted and assessed in this chapter, following a general overview on the nature of personal identity. The need for expressions of personal identity in its two forms over the Internet puts personal identity in line with the human basic needs as well as with curiosity, both of which we discussed in the two previous chapters, thus turning the Internet into an action space for the satiation of human needs in a wide sense of three angles.

Over the course of this chapter we will argue that the exposure of personal identity in its two forms over the Internet, coupled with searches for personal identities over the Web, present activities which have turned into routine performances during the short history of the Internet, but these activities have barely existed in real space prior to the introduction of the Internet. Hence, the wide-scale availability of personal identities and their search in virtual space amount to the emergence of novel human actions, both for the individuals who present their identities over the Internet as well as for their fellow identity seekers. These novel and wide-ranging Internet activities accentuate the nature of virtual space as an informational space. Furthermore, in its constitution as a novel activity performed over the Internet, personal identity differs from the satiations of human basic needs and of curiosity, which we discussed in the previous two chapters, since these two latter actions are ones which have traditionally been rooted in real space, and have been diverted, fully or partially, to virtual space, thus complementing or replacing equivalent actions in virtual space.

## The nature of personal identity

In its most basic meaning, personal identity is the conception and consciousness of the self by the self. As such, personal identity has persisted as a basic concept and as a debated issue in both philosophy and psychology. Thus, "since consciousness always accompanies thinking, and it is that which makes everyone to be what he calls self, and thereby distinguishes himself from all other thinking things: in this alone consists personal identity" (Locke 2008: 39). Such self-consciousness, can, for example, mean "the continued existence of that indivisible thing which I call *myself*. Whatever this self may be, it is something which thinks, and deliberates, and resolves, and acts, and suffers" (Reid 2008: 109). This consciousness of the self on the self is related to memory of one's past (Shoemaker 2008), and it further involves personal dimensions of continuity and sameness, taking into account changing life circumstances, as well as experiences and their imprints on the self (Hume 2008). Even more generally, then, the idea of one's personal identity involves personal awareness; personhood and persistence, all related to the self. Furthermore, personal identity involves also evidence, namely assurance to oneself of the sameness of identity under distinct and separate appearances of the self (*Stanford Encyclopedia of Philosophy* 2013). Writing just before the emergence of the Internet, Gergen (1992) argued that contemporary increasing physical and virtual mobilities have expanded social relationships and the absorption of virtual images, so that these aspects may have brought about a change in the self, turning it from an independent entity to a self of increased interdependence with others.

References by geographers to notions of personal identity have turned out to be quite rare. The contribution of geographers to the individual explorations of roots for their personal identities is in the claim that these roots may involve also some environmental contextualization, and they can, therefore, be found also around persons and not only just within them. Thus, Curry (2000: 14) argued "that privacy, whether in the form of security of place or of protection of information, is essential to the development of individual identity ... if an individual's every thought and action were open to view, the result would be chaos." Häkli and Paasi have more generally stated of the rooting of identity within the protection of information and action, or within some social context (2003: 145): "It can be argued that in the final instance individual identities are always social identities." The social roots for personal identity may be much wider, though,

since Foucault (1972) claimed, in this regard, for individual identities to constitute a product of dominant discourses with others (see also Davis (2010) for
wider social science references in this regard).

## Personal identity in social networking

Following the brief acquaintance we have made with personal identity in general,
let us turn now to personal identity as part and parcel of social networking. Social
networking has become a major human activity in virtual space, and we will
devote Chapter 8 solely to this activity. Here we would like to note specifically on
the personal identity of individuals when engaged in virtual social networking.
Social networking over the Internet can be assumed to constitute a place of con-
viviality, similarly to the ancient Agora, and in this it may act as a "third place,"
side-by-side with the place of living and the place of work (Mills 2002; Oldenburg 2000). All of these three places have contemporarily undergone some or
much of virtualization, as is evident from the discussions in the various chapters
of this volume. However, whereas the original and rather small Agora, and even
more so, the contemporary massy real malls and other physical public gathering
places, have not permitted wide networking among people who have not known
each other before hand, the Internet has opened up wide global gates for
enormous networking, notably because anonymity is acceptable, and personal
identity has, therefore, turned into a complex dimension of networking action.

The role of personal identity in virtual social networking seems to be rather
complex. On the one hand, the Internet permits the turning of identities, which
are not territorially based into something more "real" (Mills 2000), but on the
other hand, it permits, notably for adolescents, to experiment with their identity,
so that about 50 percent of them pretend over the Internet to be someone else.
This latter trend of identity pretending has been able to emerge because communications over the Internet can be performed without audio and visual cues, and
since virtual communities usually function separately from those in real life
(Valkenburg and Peter 2008). Thus, the use of *nicknames* has become common
in virtual chat rooms (Hassa 2012). The use of fake identities may, furthermore,
stem from or involve cultural dimensions, as was shown for instance for
Morocco (Hassa 2012) and Thailand (Hongladarom 2011).

The nature of online identities and their relations with offline "real" identities
has been widely debated, presenting numerous fluid and complex trends. Some
would argue that both online and offline identities are non-objective constructs,
and as we noted already "the self is something that emerges out of certain components and activities that constitute it" (Hongladarom 2011: 537). However,
these two forms of personal identity may also be viewed as turning into hybrids
with the real assuming virtual characteristics and vice versa, "so that consciousness is to some extent shared between an offline physical and an online virtual
self" (Jordan 2009: 181). Above all, it looks as if we live in "a world in which we
no longer experience a secure sense of the self" (Gergen 1991: 49). Another way
of looking at the "real" and "virtual" identities of the self is the differentiation

between the private and the public, so that "the networked self is much more public and much less private than the autonomous human subject who lies at the core of most traditional thought in the social sciences" (Warf 2013: 148).

It is not only for online and offline identities to exhibit complex interrelationships between them, but the nature of online identity per se may too be multifaceted. Online identities might be viewed as being completely anonymous, non-authentic and fluid (e.g., Turkle 1995), at other times as being stable and continuous with offline ones, while often disguising gender and race identities (Kennedy 2006), or at other times still, as changing by type of relationships and milieus (Rainie and Wellman 2012: 15). However,

> in games where we expect to play an avatar, we end up being ourselves in the most revealing ways; on social-networking sites such as Facebook, we think we will be presenting ourselves, but our profile ends up as somebody else – often the fantasy of who we want to be.
>
> (Turkle 2011: 153)

Turkle (2011: 16) went even further when claiming that "our new devices provide space for the emergence of a new state of the self, itself split between the screen and the physical real, wired into existence through technology." These complexities have made some to prefer the notion of "identification" to the more normative one of "identity" (Hall 1996). Thus, the issue of finding some evidence of personal identity is most relevant in Internet exchanges among communicating parties, frequently in attempts to gather some evidence on the sameness of one's identity, though appearing under different nicknames.

## Personal identity as information

With the Internet constituting an informational space, personal identity over it implies its being some form of manipulative and interpreted information as well: "Identity means constructing credibility, it involves negotiating a self which is coherent and meaningful to both the individual actor and the group" (Hyland 2011: 296). Thus, dealing with personal identity over the Internet has implied the construction of temporary or routine action spaces consisting of a collection of somebody's basic, verbally and visually recognizable, items which are exposed and available to others, including one's name, credentials, picture and additional items jointly comprising one's "profile." The action space constructed around somebody's personal identity as virtual information is, therefore, a double one: the posting of one's identity, and its search by others. Needless to say that identity search actions are more frequent than identity posting ones, and even more important is the fact that posting one's identity is normally done through specific and dedicated personal homepages, whereas the search for information on people is performed not only through such dedicated websites but also (and frequently even more so) through third-party sources becoming available by the use of Web search engines or via specialized websites for pay.

As peculiar as it may sound these days, the instant, wide and routine presentation and transmission of the identity of individuals, to which we have become routinely used, has been facilitated only through the Internet. Before the Internet age the sending and transmission of business cards and curriculum vitae (CVs) has been performed either manually or through the postal mail services to specific addressees, whereas the Internet permits a wide variety of identity presentations, accompanied by a variety of access channels to them, either directly by individuals or indirectly by specializing websites. Identities may, thus, be presented and exchanged through emails, through personal or company websites, or through pages on social networks. As Dominick (1999: 647) pointed out: "Prior to web pages, only the privileged – celebrities, politicians, media magnets, advertisers – had access to the mass audience." As such, the Web has become a one-to-many communications medium for the presentation of personal identity (Miller and Mather 1998). This democratization of the posting, or the self-provision to others, of personal identity information has emerged also for people's search for personal identity information: before the introduction of the Web only the privileged ones who paid significant sums of money to private investigators were able to collect information on individuals, information which can now be accessed by all Internet users within seconds. The Web has, thus, become also a many-to-one (or many-to-many) communications medium regarding the search for people's personal identities. In other words, the Web has facilitated the emergence of a two-way action space for the posting of as well as for the search for personal identities.

Basic personal identities as revealed to fellow individuals in the real world are different in principle from those found in the virtual one, in that identities in the real world are foremost physical: face; body; voice and body language, accompanied by some verbal information possibly becoming available on the person, whereas in the virtual world the order is exactly the opposite: verbal information is the basic norm, and visual and oral ones might sometimes be available, as well. However, this seemingly major difference between the two spaces has been blurred in that pictures and even short movies become now more frequently available on personal Web pages, side-by-side with URL (uniform resource locator) addresses of personal Web pages appearing on printed personal cards handed out in real space.

The major virtual framework or channel for presentations of personal identity by individuals on the Web is the personal homepage. No estimates could be found as to the number of such pages, but it was believed at the time to be a widely used tool (Papacharissi 2002). It has become customary for personal homepages to contain the following items: self-description; childhood and youth memories; family and friends; pets; houses and places of residence; cars and bikes; trips; free time activities; important personal events and dates, all being items which are valued as socially and culturally positive (Glatzmeier and Steinhardt 2005).

Goffman (1959) studied at the time the presentation of the self in everyday life, mainly within the prevailing framework for such presentations at the time, namely face-to-face encounters. He differentiated between the impressions one gives and those that are given off, or put another way, between overt and covert impressions, respectively (see Papacharissi 2002; Davis 2010). Personal homepages permit the simultaneous transmission of both impression classes via texts (overt), as well as through pictures, animations and links (covert). An important difference between overt and covert impressions in real as compared to virtual spaces is that Internet self-presentations can be controlled, whereas body language in real space is largely uncontrolled. However, portals offered by ISPs for personal homepages may restrict quantitatively the materials individuals may wish to put on their personal pages (Papacharissi 2002). Still, some flexibility may be provided in this regard by the use of hypertexts that permit a non-hierarchical structure of the presented materials. This is also the case for the preferred framework for personal pages of content-dedicated rooms within a wider website (Miller and Mather 1998).

The growing use of personal homepages for the exposure and presentation of personal identities may bring about some personal "fights" for attention by others. "Attention becomes a good that is in short supply and has to be fought for using self-presentation" (Glatzmeier and Steinhardt 2005: 33; see also Franck 1998). The personal homepage has, thus, to initiate some curiosity at the browsing person, either through graphics or through some inviting comments, side-by-side with keeping some common modesty by the page owner.

Personal homepages may be differentiated by owners' age as well as by other social and economic parameters of their owners, but a special kind of personal homepages are the academic ones, prepared by university faculty, and recently explored by Hyland (2011). Since personal pages of faculty are posted within websites of the universities with which the faculty are affiliated, the pages reflect not only self-presentation motives of the faculty but also the university efforts for the attainment of prestige. Therefore, this restricts the posting of non-academic personal information by the faculty. Like other personal homepages, those of faculty permit continuous changes to be made on them, a quality that is of specific importance for faculty who are involved in the publication of research and who are further interested specifically in the search of up-to-date information on research publications by their colleagues. Thus, it was estimated for 2007 that some 71 percent of university faculty had a personal homepage (Tang *et al.* 2007). The academic personal homepage constitutes mostly a digital version of the printed CV, frequently enhanced by a personal picture and/or links to their publications which are available online. The major change from print to digital in this case is not the very documentation of personal identity but rather its global distribution and availability to all interested parties without seeking prior permission from the posting faculty for its use, as has been the case with printed CVs. This change implies a most significant transition in the work routines of both faculty and academic institutions, making it possible for them to learn widely on the work and products of colleagues and potential partners and employees elsewhere.

## Search for personal information

Our attention in the previous section was focused on the supply of personal identity information over the Internet. We will now attempt to explore the demand side, or the search for personal identity information, over the Internet. Individuals may engage in the search of information on fellow humans for a variety of reasons: they might wish to assess the professional qualifications of people, notably if their services might be ordered, or these people may be hired; hearing about somebody socially might lead to intensive information searches, notably if dating is involved; and finally, people might wish to learn about others just out of sheer curiosity.

The virtual action space for people searching for information on others is most diverse and complex, and with the use of search engines covering the whole Web, including or excluding Facebook and other Web 2.0 networking systems, it becomes truly global. Rainie and Wellman (2012) termed such wide virtual searches as "coveillance," and we will refer to the negative sides of such searches in Chapter 9. As previously mentioned, such searches for personal information on others would not have been possible through real space facilities, such as libraries, nor through the use of older communications technologies, such as the telephone, which normally requires prior acquaintance between communicating parties, in which one partner is seeking some information, and it would yield only partial and sometimes problematic information. The advantage of the Web as an informational action space is, therefore, unparalleled. Thus, for 2012 it was found that 51 percent of American adult Internet users searched for information about someone whom they knew already or on somebody whom they might potentially meet (PewInternet 2013a).

The search for information on people who are not subscribers of social networks may be performed for many people but not for all. Thus, one may easily find personal information on professionals, and more so in developed countries. Obviously the search would begin with personal homepages, if available, but one of the major advantages of the Web is that it permits searching also third-party sources on people. This advantage is such for searching parties, whereas for searched people it can be an advantage or a disadvantage depending on the contents and meaning of the materials available on them on the Web. Moreover, such third-party materials may have an impact on one's life in real space, and they may have a lasting such impact if they remain on the Web without deletion (Schmidt and Cohen 2013).

Another advantage of the patterns of information organization and access in the Internet is that they permit simultaneously both systematic and associative searches. Systematic searches become possible mainly through the reading of one's personal homepage, through one's workplace websites and data, and through information search companies, whereas associative searches are made possible through the variety of sources offered by search engines, and through the use of hypertext, which automatically moves the searcher to completely different, but potentially relevant, sources.

Attempting to construct a profile of one's personal identity based on collected materials from the Web might imply coping with the trustworthiness of the various materials, as well as the coping with contradicting materials. The assessments of information pieces may require the use of professional services offered by information specialists in real or virtual space, notably if the collection of materials is performed within the framework of personnel hiring. Thus, the wealth of information on individuals over the Internet may not only imply vulnerability for the individuals on whom information is gathered, but it may further pose a challenge of information processing and judgment for the searchers.

## Conclusion

Table 6.1 attempts to summarize the classifications of personal identity that have been offered in this chapter. We indicated two actions regarding personal identities performed over the Internet: presentation of identities, or the supply side of identities, and the search for identities, or the demand side. People may present some details of their internal identity through personal homepages or over social networks, and they might present their external-professional identity through the posting of their CVs and/or personal homepages. Internet users are, furthermore, frequently involved in fake and pretended presentations of themselves, notably in social networks. Presenters of identity, as well as non-presenters, may often look for information on others, either for professional, social or personal motives. Such searches are performed through dedicated websites for people searches, which normally provide paid-for services, and they may make use of search engines.

Despite the obvious differences between real and virtual spaces, there are numerous and rather basic characteristics and complexities of personal identity which prevail in both spaces: first, in both spaces people may present several and rather changing identities; second, in both spaces people may present "true" and fake identities; third, in both spaces there are identities presented by individuals, side-by-side with their identities as presented by others; fourth, in both spaces people present their given/overt identities, complemented by their given-off/ covert personal identities as perceived by fellow people; and, fifth, in both spaces persons may possess a more stable identity as presented by their CVs in real space and by the equivalent personal homepages in virtual one, side-by-side with changing identities as presented through face-to-face and online chats.

*Table 6.1* Personal identity: needs and virtual expressions

| Action | Need | Virtual expression |
| --- | --- | --- |
| Presentation | Internal identity | Personal homepage; social networking |
| | External-professional identity | CV; Personal homepage |
| | Pretended identity | Social networking |
| Search | Professional or social | People's search websites; search engines |
| | Personal curiosity | |

The major element which exists only offline in real space and cannot exist by definition in virtual space is one's internal personal identity, though this internal conception and consciousness of the self may be influenced, and may be even changed, by one's own personal identity as presented over the Internet by oneself and by others, since the internal identity is generally sensitive to contextual-social circumstances.

The major dimensions of online personal identities reflect the unique nature of the relatively novel informational virtual action space: first, the diffusion of the personal identities of individuals has become potentially global, assuming that everybody's personal homepage maybe accessed by everybody else located anywhere; second, and the other side of this coin, people's personal identities can be accessed by everybody else without prior permission granted by the owners of the personal homepages; third, the availability to others of "third party" materials relating to one's personal identity is enormous and easily accessible, as compared to equivalent availabilities in real space; and fourth, the total quantity of personal information on people over the Internet might be most substantial. These characteristics of online personal identity channels imply that much power has been granted to individuals in their attempts to present them-selves, even if these attempts are competitive in some way, and by the same token, even more power has been granted to individuals in their seeking of personal identity information on others.

# Part III

# The Internet as an action space for individuals

# 7   Daily activities

The lines between information, communication, and action have blurred: Networked individuals use the Internet, mobile phones, and social networks to get information at their fingertips and act on it, empowering their claim to expertise (whether valid or not).

(Rainie and Wellman 2012: 14)

Following the second part of this book, which focused on human basic needs and their satiation through the Internet, this final part of the book will outline and discuss some major daily activities performed through the Web, such as shopping, learning and the attainment of governmental services; social relations as established, fostered and maintained via online social networking; and finally some darker human actions performed through the Internet. This quite diversified list of human virtual actions does not pretend to be exclusive, so that the purpose of the following three chapters is mainly to present some major and rather varied routine human activities in the virtual world, and the way the Web serves as action space in this regard. In addition to e-activities performed by individuals there are other actions which are executed by professionals and institutions on behalf of individuals, such as e-health, which includes remote medical testing and consultations, and we will devote some attention to it as well, later in this chapter.

There are additional individual routine activities, of a more cultural nature, for which the Internet is not the sole alternative to real world attendance. Virtual alternatives to the theater and the concert hall have been there already before the introduction of the Internet through the television and audio home systems, as well as the more contemporary MP3 devices, and mostly people have preferred to use these latter options rather than the Internet for such activities. At yet another arena, that of religious services, the Internet may offer an alternative to churches and temples, offering even online religious communities (see, e.g., Kogan 2001).

The seven major "*e-*" (standing for electronic) activities by individuals, which will be presented in this chapter are: home-based work; shopping online; e-banking; e-learning; e-government; travel online and e-health. In exploring these activities, we will outline online functions that are widely used by individuals, side-by-side with some rather restricted operations by individuals in several of these online activities. As far as possible, data will be presented on the percentage of people acting online in each of these activities, either partially, or wherever possible also fully, such as the percentages of workers who work only online, the percentage of customers shopping online; and even some data on students who study online. The special nature of the relationships between physical and virtual spaces for daily activities will also be highlighted, pointing to complementarity between the two action spaces rather than to substitution of the real by the virtual. We will begin our explorations with two general discussions, the first of which will highlight daily human actions as such, and it will be followed by a second one, presenting a brief exploration of broadband communications and their impact on daily online actions. These two general discussions will be followed by explorations of the seven main human actions and their action spaces over the Internet. We will conclude these particular presentations of virtual actions with some comparisons among them.

## Daily virtual actions

The contemporary extended availability of access to the Internet, notably through personal fixed and mobile broadband, can be interpreted as providing enhanced locational opportunities of action for individuals, thus implying their self-extension. Communicating actors can reach people, information, or virtual destinations and services, equally from wherever they are located, or from any point of communications origin. But even more importantly, from the perspective of locational opportunities, is the distanciation of human individual action (Giddens 1990), which we mentioned already. It refers to the increasing geographical spread of potential destinations for the retrieval and sending of information, and the ability of people to work as well as to buy services globally, thus widening their locational opportunities and their action spaces. Cyberspace has not merely emerged as a new world of information, but it has further evolved into a dense concept of converging technological environments, human minds, personal motivations, and into a source for the generation of all kinds of human artifacts (Dodge and Kitchin 2001; Kellerman and Paradiso 2007; see also Part 1).

The contemporary scene of the information age presents to individuals a wide choice among media for a variety of performances via virtual action space, including mainly PCs, laptops, tablets and smartphones. It has become possible, therefore, for a person on the road to communicate with information sources available through the Web, while these sources may simultaneously change (e.g., through peering) their virtual locations (or hosting) as the using person moves, still without any ramifications to the user's ability to access these websites when the locations of these information sources change. In other words, new sophisticated virtual mobilities permit us to act virtually without any regard to the location of the cities or countries from which we communicate, nor with regard to the internal spatial structures of cities from which we communicate.

The operation of contemporary personal communications media by their users, notably fixed and mobile telephones and the Internet, is assumed to be rather convenient and user-friendly, and thus barrier-free. Once communications is initiated, the speed of reach of other places is normally instant. Moreover, users of communications media expect constant enhancements of communications speed through IT innovations or upgraded infrastructures. The growing convenience and speed of communications have resulted in the expansion of the extensibility and accessibility of users, making it possible for them to reach eventually either new or veteran opportunities (see Janelle 1973; Adams 1995; Kwan 2001). Repeated uses of such virtual opportunities may potentially, at least, change the balance between services offered through facilities located in real space and within easy physical reach of potential customers, on the one hand, and virtual services provided solely through the Internet, or services located anywhere in real space and reached through the Internet, on the other. Such possible changes would obviously be in favor of the virtual and at the expense of the real, or they may potentially bring about some structural changes in business facilities located in real space, making them more attractive to potential customers.

## Broadband interaction for daily uses

"Broadband has become the standard for Internet use" (Mossberger *et al.* 2013: 3). The global traffic of mobile data has doubled every year since 2008 (Allot 2013), and a significant portion of this traffic may stem from daily actions by users, other than for entertainment, which has turned out to be the most dominant producer of heavy traffic, given the volume of music and video clips (Kellerman 2010). Thus, it was estimated that in 2012 some 88 percent of smartphone users in the US had travel reservation applications installed in their devices, followed by 79 percent of them having shopping applications installed in their smartphones (*The Economist* 2012). By the same token, in 2012 some 80 percent of US smartphone owners accessed retail content through their devices (ComScore 2012).

The introduction and wide adoption of mobile broadband implies instant access to cyberspace for its users, but it may further carry implications for the use and meaning of urban physical space for smartphone users. The constant availability of GPS, Google Maps and LBS, even while walking or driving through an unknown city, or through unknown parts of a known city, implies an efficient moving of people through urban space when aiming to reach specific addresses, thus saving time and efforts in walking, driving and searching. It further implies a live and ongoing integration of virtual and real action spaces, with the virtual one guiding physical movements within real space. However, this efficient crossing of real space with the aid of tools located in virtual action space turns the crossed streets and urban space at large into a kind of impediment rather than into an occasion for some cultural exploration. Hence, this specific integration of virtual and real action spaces reduces one's urban experience, in that cities turn into a mere mosaic of places of production and consumption, ignoring the traditional role of cities as providing residents and visitors with passive or active experiences of human life at large, such as through the encounter with city rhythms at different times of the day and the week (see Allen 1999).

This loss of meaning of urban space for constant GPS users may be tied to a potentially wider possible loss of contact with the immediate physical environment by mobile broadband users, due to other uses of the virtual action space in which they may be engaged through their smartphones while moving about urban public sphere, notably their global communications for social networking, their consumption of virtual services, and their obtaining of information. An additional front of blurring among spaces is the decline in the traditional separation between home and work activities which was recognized when access to the Internet was made first through fixed PCs only (see Kellerman 2006, 2012). However, there is, of course, another side to the coin of transitions in lieu of the growing use of mobile access to the Internet, which has been enhanced by the ease of use of smartphone and tablet versions of browsers and website applications. Portable Internet devices have made it possible for users to consume services without limitations of time and location in their routine daily life. This ease of universal access to services has brought about a vicious cycle with real space facilities, amplifying even further the blurring of separation of the temporal and locational dimensions in the consumption of daily activities, as well as the separation between real and virtual action spaces.

## Home-based work

We will turn now to discussions of seven major "*e-*" activities, the first of which will be home-based work. Home-based work, or as it has variously been termed also as *telework, telecommuting* or *telehomework*, assumes that people's work can be performed by many workers from home, using software consisting of the Internet and specialized software programs and packages if needed, operated through a PC or laptop, equipped with a web camera, side-by-side with a telephone and fax (which might be integrated also within PCs and laptops).

Home-based work has probably been the first and continuously discussed action space and potential locational opportunity for individuals equipped with the needed ICT devices (see Kellerman 1984). For Halford (2005), work at home implies the spatial *relocation* of work coupled with its *dislocation* into cyberspace. Work at home, fully or partially, became potentially possible as far back as the 1970s, following the diffusion of the *Telnet* technology for remote access to mainframe computers, using computer terminals and later on PCs. The introduction of the Internet and its extremely wide options for information transmission some 20 years later, in the mid-1990s, could have potentially turned work from home into a standard form of work, notably following the previously accumulated 20-year experience of the pre-Internet computerized work options from home. At least, well-equipped homes should have become potential part-time work places, using virtual action space.

As things have turned out, however, work at home as an exclusive location for work has not turned into a favorite option, neither for employers, nor for employees or for the self-employed. Employers have preferred their on-site supervision of workers, as well as the holding of face-to-face staff meetings on a regular basis, and employees' on their part, preferred to meet face-to-face with their colleagues, on a daily or almost daily basis, and be involved with goings-on in their offices. Women in particular have tended to prefer a geographical separation between home and work duties (see Halford 2005; Hislop and Axtell 2007; Felstead *et al.* 2005; and Loo 2012 for reviews). Still, though, partial work from home, notably after work hours, has become routine for many workers, who sometimes have turned into so-called "nomads" (see, e.g., Castells 2000). As such, recent Finnish findings have shown that telework is more common among knowledge-intensive workers (Merisalo *et al.* 2013). Work at home has turned out not to imply higher quality of life for workers, although productivity was reported to be higher for workers from home (Halford 2005).

Home-based workers may be classified into three groups: partial or hybrid; full-time and "mobile workers," those who work at dispersed places such as at clients' facilities (see Malecki and Moriset 2008). Of these three groups, people who work wholly from home are of special interest, since such workers are supposed to have established their work fully online, so that their daily professional action space has been fully virtualized. Even more interesting are the self-employed workers who work fully from home, mobilizing and serving their clients through the Internet, and performing their work through the Internet as well, for instance text editors and translators or consultants.

Hence, let us examine now some data on home-based work in several countries worldwide. For the year 2000, a low rate of just 0.6 percent was estimated for employees who worked wholly from home, out of the total employees in European countries (Bates and Huws 2002; see also Helminen and Ristimäki 2007). In 2005 the percentage of those involved in telework "almost all of the time" in the EU-27 countries rose to 1.7, led by the Czech Republic with 9.0 percent, side-by-side with some additional 7.0 percent of the EU workers who were involved in telework at least "a quarter of the time" (Welz and Wolf 2010). For the UK, in particular, in 2002 it was estimated that 2.4 percent of the UK workers worked mainly, but not fully, from home (Hislop and Axtell 2007), rising to some 2.5 percent in 2005 (Welz and Wolf 2010), and rising again to some 4.9 percent in 2010 (Telework Association 2012).

In the US, which was the pioneering society in the adoption of computers for both offices and homes, in 2008 only 1 percent of American employers allowed all or most of their employees to work some regular paid hours at home on a regular base (US Bureau of the Census 2010). In 2005 some 5.6 percent of the US workers worked exclusively from home, rising to 6.6 percent in 2010 (US Bureau of the Census 2012b). However these percentages included farmers, construction workers and manufacturers, considered traditionally as working from home. Thus, in 2005 only some 1.7 percent of US workers performed professional and related services from home (see US Bureau of the Census SIPP

2006b). In 2010 nearly one half of the home-based workers in the US were self-employed (US Bureau of the Census 2012b).

The trends for home-based work which we have just observed for both the US and the EU point to constant growth in the numbers and percentages of those working from home, including those who work only or primarily from home, but that this pattern of location for work is still modest, consisting of single-figure percentages of the workforce. In other words, only relatively few people in countries which have led in the personal adoption of Internet connectivity are ready to use the virtual action space as the sole action space for their work, whereas the vast majority of workers still prefer a mix of physical and real action spaces for their daily work space. Face-to-face work and work supervision are still highly valued by employees and employers, respectively.

## Online shopping

In our discussion of online shopping we will focus on what for merchants is called B2C (business to customers) e-commerce, as compared to B2B (business to business) e-commerce (see, e.g., Loo 2012). B2C is usually termed by customers as *online shopping* and this is the term that we will refer to below. We will attempt to explore the virtual shopping action space, which implies a possible emergence of geographical opportunities for individual consumers to shop beyond their normal daily physical reach vis-à-vis online shopping. We will do so for two potential geographical levels of online shopping: domestic and international. The data which we will present and interpret are mostly at the national level, as systematic data on online shopping at the regional and local scales, do not seem to be available as of yet. Furthermore, our discussion here will focus on "shopping" in its rather traditional sense, referring to the purchase of products and not to the purchase of services, such as flight and hotel reservations, which are increasingly made through the Internet, and which we will discuss separately in a later section as "travel online."

The virtual action space for shopping permits potential customers to perform virtually all the phases which are included in shopping in a real space store, except for touching and trying: search for products; search for vendors; viewing of products and reading of information on them; price comparisons; and eventually the very act of purchasing. Shopping has, thus, become geographically extended into a potential global "opportunity" when shopping through global stores and shopping mall, accessed via virtual action space. Moreover, this global virtual marketplace functions continuously, on a 24/7 basis (Loo 2012), so that online shopping frees customers not only from spatial barriers but also from temporal frictions and restrictions. On the other hand, one has to bear in mind the *fragmentation of activity*, as a common characteristic of shopping activities of individuals, calling for some, and if only basic, continued physical shopping, side-by-side with virtual one (Couclelis 2004; see also Schwanen *et al.* 2008). Thus, information products, "soft" ones such as reports, music and books, side-by-side with "hard" IT products, such as computers of all types, have tended, in the early 2000s, to be popular for sale online, whereas daily food shopping, as well as less-frequent car purchasing, are mostly done offline (US Bureau of the Census 2003).

As we mentioned already, online shopping, by its very nature, does not permit the touching and trying of merchandise, but side-by-side with this constraint, it permits extremely wide and instant price comparison, which are unparalleled in real space. It further provides for plenty of information on products. Another feature of online shopping is that it is also a major example of the availability of a silent option for the very performance of shopping, as compared to noisy shopping malls in real space. As we noted already in Chapter 1, noise is widely used these days as a marketing tool. Mass consumption has emerged through shopping malls, but the same type of consumption may now be fully or partially performed through the Internet. We noted before that the noisy mall option permits touching and testing of products, whereas the silent one permits extensive price comparisons, but consumers may frequently engage in both options, notably when it comes to the purchase of major appliances and other highly priced commodities, eventually making the very purchase eventually either in a physical store or in a virtual one.

Following the general discussions of shopping online it is time to ask: do people actually make use of the global opportunity and action space for shopping without any geographical restrictions of physical place and distance, or do they still prefer to stick to the traditional way of shopping in physical action space rather using corporeally accessible stores? This question is even more interesting when online shopping is divided into domestically-based shopping websites ("closer" ones, geographically and culturally) and foreign, more "remote" ones.

As we just noted, for many contemporary customers a given shopping activity may constitute a double experience, involving a mixture of online components and physical ones. Thus, one may browse online for products, makers, sellers and prices and eventually shop in a physical store or vice versa. The affordance of the Internet for e-commerce is, therefore, in many cases, partial (Schwanen *et al.* 2008). In the following examination of data for shopping online, we will focus on the very final act of online purchasing, as this concluding phase of shopping implies also domestic or international transfer of funds, as well as remote domestic or international guarantee by sellers. Our focus will be less on the aspects of financial volume of spending by individuals, and the financial extent of B2C e-commerce, and other related aspects (see, e.g., Kellerman 2002: 127–33; Visser and Lanzendorf 2004). Rather, we will concentrate on the behavior of customers, expressed in the percentage population shopping online, in order to see whether the very availability of a virtual shopping action space, which poses an opportunity to shop indifferently from a locational perspective, is really materialized, or alternatively the physical action space for shopping in physical stores still governs shopping behavior.

As far as online shopping in general (domestically and in foreign countries) is concerned, measured by the percentage of the total population shopping online, it seems that the use of this medium has become quite popular in North America but less so in Europe. Thus, already by 2003 some 40 percent of Canadian households made online purchases. However, Ontario accounted for almost one half of total online spending (Statistics Canada 2004). In the US, in 2001, some 21 percent of the total population purchased online (NTIA 2002), rising to 71 percent of adults a decade later in 2011! (PewInternet 2013a). However, the share of e-retail sales out of the total retail sales in 2001 was only 1.1 percent, rising to 2.4 percent in 2005 (Malecki and Moriset 2008), to 3.6 percent in 2008 (US Bureau of the Census 2011), and to 4.0 percent in 2009 (US Bureau of the Census 2012a). Thus, similarly to home-based work, online retail sales are growing, and contrary to home-based work, shopping online has become highly popular, but still the vast share of shopping is still done physically, though possibly aided by online tools, such as price comparison and store choice.

The scene for online shopping in the EU has been more complex. Generally in the EU, in 2005–2006 only some 23 percent of the total population did so, and this percentage was equal to the share of the population purchasing through the more traditional channel for shopping at a distance, the postal service! (European Commission 2006). Four years later, in 2010, this rate increased to 33 percent, growing remarkably further to 59 percent in 2012 (Eurostat 2012a). The EU, though, has been highly diversified with regard to the tendency for online shopping, including countries with high population percentages of domestic online shopping in 2010, led by Sweden (54 percent), the Netherlands (52 percent), and Germany (47 percent), and going all the way down to 7 percent in Malta. Previous data for 2005–2006 present, obviously, more modest rates (Table 7.1).

Purchasing abroad may be considered as an advanced mode of domestic online shopping, since it may involve money exchange, trusting a foreign company, as well as overcoming an image of remoteness. Here too, there was found a major difference between North America and the EU. Whereas in Canada in 2003, one-third of online purchases were made through foreign websites (Statistics Canada 2004), probably mainly American ones, in the EU in 2005–2006, the percentage of purchases made in other EU countries stood at only 6 percent, increasing only modestly to 7 percent in 2010, and to 10 percent in 2012 (Eurostat 2012a, see also Aoyama 2003). This rather low rate of international online purchases within the EU has emerged despite the convenience of possibly using the same currency, the Euro, in most purchases. The growth trend for cross-border online shopping in the EU between 2002 and 2006 was assessed by the EU at the time as being generally high, with several countries doubling the percentage of population using cross-border online shopping (European Commission 2006). However, later on, between 2006 and 2010, alongside impressive growth in cross-border shopping in small countries (e.g., Ireland, Cyprus and Malta), other countries presented slight declines (e.g., Netherlands, Sweden and Belgium).

The trends for EU-wide online shopping may be viewed also another way, when taking a look at the rates of Internet adoption. The EU percentage of

*Table 7.1* Percentage population in EU countries shopping online in EU countries and/or domestically, 2005–2006 and 2010

| Country | Shopping in EU countries 2005–2006 | Shopping domestically 2005–2006 | Shopping in EU countries 2010 | Shopping domestically 2010 |
|---|---|---|---|---|
| Luxembourg | 28 | 7 | 38 | 12 |
| Denmark | 19 | 46 | 24 | 32 |
| Austria | 18 | 21 | 30 | 32 |
| Netherlands | 15 | 46 | 12 | 52 |
| Sweden | 14 | 45 | 13 | 54 |
| Finland | 13 | 34 | 18 | 38 |
| Ireland | 12 | 19 | 34 | 26 |
| Belgium | 12 | 17 | 11 | 19 |
| Malta | 11 | 4 | 39 | 7 |
| UK | 7 | 41 | 9 | 53 |
| France | 7 | 26 | 9 | 38 |
| Germany | 4 | 30 | 6 | 47 |
| Estonia | 4 | 17 | 8 | 21 |
| Italy | 4 | 11 | 4 | 13 |
| Slovenia | 4 | 10 | 8 | 23 |
| Spain | 4 | 9 | 6 | 21 |
| Cyprus | 4 | 0 | 22 | 11 |
| Czech Republic | 3 | 21 | 4 | 38 |
| Latvia | 3 | 11 | 7 | 17 |
| Poland | 2 | 16 | 3 | 35 |
| Portugal | 2 | 3 | 5 | 12 |
| Greece | 1 | 2 | 8 | 12 |
| Slovakia | 1 | 5 | 18 | 30 |
| Hungary | 1 | 8 | 4 | 23 |
| Lithuania | 1 | 4 | 7 | 17 |
| EU general | 6 | 23 | 7 | 33 |

Sources: Table: Kellerman 2012b, Table 11.1. Data: European Commission, 2006 (illustrations D46 and QB1.1) and 2011 (p. 15).

homes connected to the Internet grew from the 2005 rate of 44 percent to 73 percent in 2012, or a growth rate of some 66 percent (Eurostat 2012a), and in parallel the percentage of those who purchased abroad through the Internet grew, as we noted, from 6 percent in 2005–2006 to 10 percent in 2012, or a growth of some 66.6 percent. In other words, the expansion of Internet users and the expansion of Internet use via cross-border online shopping have been similar by their growth rates from 2005 to 2012. The modest change in cross-border online shopping was coupled, however, with some modest growth in the much higher rates of domestic online shopping, increasing from 23 percent of the population shopping online in 2005–2006 to 30 percent in 2008, and again to 33 percent in 2010. The modest growth in international online shopping within the EU attests to the still existing apprehension of international shopping online, even in other EU countries (Commission of the European Communities 2009, European Commission 2011).

Two factors seem to govern the European online shopping behavior. First, country size matters. Citizens of small countries tend to engage more in foreign shopping than those of larger ones. Thus, Luxembourg led the list of foreign shopping in 2006 with a percentage (28 percent) that was four times higher than its percentage for domestic online shopping (7 percent), and Malta performed similarly in 2010. Cyprus presented a similar pattern with no domestic online shopping at all in 2006, whereas the Maltese population tended over 5.5 times more to purchase internationally than domestically in 2010 (39 percent and 7 percent respectively).

The opposite trend was true for three of the larger European countries, which presented more domestic online shopping than foreign ones: UK (41 percent for domestic shopping versus 7 percent abroad in 2006, and 53 and 9 percents respectively for 2010); France (26 percent and 7 percent for 2006, and 38 and 9 for 2010 respectively); and Germany (30 percent and 4 percent for 2006, and 47 and 6 percent for 2010 respectively). This preference for domestic online shopping in bigger countries is based on the more extensive online action spaces for domestic shopping which is available in them, as compared to those in smaller ones, in which the very availability of foreign online shopping is attractive, since they enjoy a more restricted availability of domestic online shopping. The importance of country size is of interest since previous studies of international B2C e-commerce identified other key factors for its magnitude, mainly Internet infrastructure, economic development and culture (Hwang *et al.* 2006; Kshetri 2001).

Second to country size for the explanation of differences in the volume of international online shopping in the EU might be national Internet penetration rates. The Scandinavian countries have led the adoption of the Internet, following their leadership in the adoption of telecommunications media at large since the introduction of the telephone, back in the nineteenth century (Kellerman 1999). This tendency may have led to higher foreign and domestic online shopping. The Internet has, thus, brought about some change in the shopping habits of Scandinavians, as well as in the shopping patterns of residents in other small and well-developed EU countries.

Asian trends for shopping online have presented global leadership as far as online shopping is concerned, notably for shopping via smartphones. Thus, in late 2012 some 55 percent of the Internet users in China made a purchase using a smartphone, and most major countries surpassed the US rate of 19 percent and the Canadian rate of 13 percent for shopping through smartphones in late 2012 (Table 7.2) (European Travel Commission 2013). This leadership in the usage of mobile broadband applications continues the global predominance of the region in the penetration of broadband at large, and of mobile broadband in particular (see Kellerman 2006b, 2010).

The overall picture of the quite modest shares of the population who make use of online shopping, mainly in Europe, seems to suggest that the geographical location of shoppers, coupled with their preference for shopping in traditional stores located in real space, are still leading factors in the choice of shopping

*Table 7.2* Percent Internet users purchasing via smartphones, Asia and the Pacific late
2012

| Country | Percentage purchasers |
|---------|----------------------|
| China | 55 |
| South Korea | 37 |
| India | 26 |
| Indonesia | 26 |
| Vietnam | 24 |
| Malaysia | 23 |
| Thailand | 23 |
| Japan | 22 |
| Australia | 17 |
| Philippines | 15 |

Data source: European Travel Commission (2013).

action space, and these factors might well continue to be of significance in the future, as even the North American data show. This tendency of individuals for the preferred use of real action space is still present due to shoppers' attitudes: habits, lower trust in online vendors, and uncertainties about online financial transactions were suggested as the main obstacles on the road of expanding online shopping (European Commission 2002). Furthermore, real space shopping permits the touching and the trying on of merchandise, and this turns out to be of importance to many customers, so that the very action of purchasing tends to constitute also a kind of entertainment, and thus slowing down the pace of a wide adoption of online shopping.

Looking towards the future, e-commerce may turn into a form of empowerment for women, since it facilitates shopping in privacy from home, and this might be of significance notably in traditional societies in developing countries when economic conditions will permit a significant penetration of online shopping. On the supply side, merchants have to invest more in websites aimed at foreign countries in order to attract customers from there, targeting mainly the younger generations for whom cross-border activities at large might be mentally easier. Needless to say that websites have to be multilingual in order to facilitate international shopping, notably if there emerges some competition between international and domestic shopping websites.

## E-government

In her recent geographical review, Loo (2012: 37) defined e-government as: "attempts of governments to automate and facilitate the use of e-technologies in governmental processes, whereby individuals and organizations interact directly with governments for various reasons" (see also Yildiz 2007). Warf (2013) reviewed too the recent geographies of e-government from slightly different angles, and he differentiated among several categories of e-governments, similarly to those of e-commerce: government to business (G2B); government to government (G2G) and government to citizens (G2C). Given our focus in this volume on virtual action spaces for individuals, our concern here is mostly, of course, with G2C. From a governmental perspective, or the supply side of e-government services to citizens, the development of e-government to citizens involves efforts for the modernization, higher efficiency and better services to the citizens–clientele. This latter approach to e-government systems has become a most popular one among governments, and it was termed "managerial."

There are, however, additional approaches to the virtual relations between governments and their citizenries, notably: the offering of voting services ("consultative"), and an offering of services and public opinion channels jointly with other role players ("participatory") (Chadwick and Nay 2003; Warf 2013). The development of managerial services may present a hierarchical evolution from the bottom to the top, starting with a one-way presentation of services by governments online, and culminating in a full interaction with citizens, including fee payment and the virtual transmission of documentation, which may occur following the computerized integration of inputs by various governmental levels (Layne and Lee 2001). Typical obstacles in the road to full e-government services include: authentication of citizens; mobilizing the bureaucracy for change in service structure and means, and computer capabilities of the citizenry (Loo 2012).

We may identify a major difference between G2C online services, on the one hand, and the numerous other "e-" activities, outlined in this chapter, such as shopping, learning, banking and travel arrangements, on the other. E-government relates to a compulsory and uncompetitive service. One may choose what to buy and where to buy, which is true also for the attainment of other rather commercial services, such as banking and travel. However, one may be forced to obtain governmental services, such as payment of taxes and fees, submissions of applications for permits, and the receipt of documentation. This compelled system of services applies to all governmental levels: local, regional and national. Thus, one can choose neither among governments, nor among suppliers of governmental services and their virtual offerings. If citizens wish to change their suppliers of e-government services, they have to move to a new residential location, but still even in the new location one may find again a similar uncompetitive and compulsory attainment of services. The other side of this coin of this compulsory and uncompetitive scene is that for many citizens in any country the possibility to interact with governments electronically, mainly via the Internet, might imply a major advantage as compared to that of interacting with commercial service suppliers, since digital interactions with governments involve a major time saving as compared to the attainment of traditional face-to-face services, and the automatic systems may further turn out as friendly ones.

The geographical reviews of e-government by Loo (2012) and Warf (2013) point further to a familiar international landscape of a digital gap in the emergence of G2C online services: they are rapidly developing in developed countries while being still in their infancy in developing countries, with some exceptions among developed countries, notably the reluctance of Japan with regard to e-government services. This digital gap is amplified when it comes to e-government services, because the attainment of these services requires sufficient literacy by the clients-citizens. The international gap in the provision of e-government services is further coupled with intranational ones, with lower potential uses among the poor and the elderly.

Our interest here is not just with the very availability of governmental virtual action spaces for individuals but also with their use by the citizens, in cases where e-government services are offered to citizens in ample or reasonable supply. The use of managerial e-governmental services by citizens might be either "passive" in the sense that citizens merely access information on governmental websites, or it might be "active," so that citizens interact with their governments by sending information to governments. Thus, for 2009 the EU reported that some 28 percent of the European citizens accessed information on governmental websites, as "passive" users, whereas only 13 percent, or less than half of those who accessed governmental information online, sent information electronically to governments, or acted as "active" users, by then. By the same token, PewInternet (2013a) reported for the US that in 2011 some 67 percent of adult Americans visited a local, state or governmental website, but for a year earlier, 2010, the OECD (Organization for Economic Cooperation and Development) (2012b) reported that only 41 percent of Americans "used" e-government services.

Table 7.3 reports what seems to constitute the two-way usage and not a mere access of citizens in OECD member countries to e-government services in 2005 and 2010. The average for the 26 OECD member countries was 28 percent for 2006 rising to 42 percent in 2010. In commenting on these data the OECD (2012a: 1) remarked that,

> citizens' uptake of e-government services remains lower than expected even in the best performing countries.... One possible explanation is that vulnerable segments of society are unable to utilize digital channels due to lack of awareness or IT skills. Another is that the online services offered are not always responsive to individuals' needs.

Lack of trust by citizens in governmental e-services might be added to the first explanation relating to lack of access and awareness to the services, whereas the lack of competition might be added to the second one relating to some possible lack of responsiveness of e-government services.

The data for specific countries point again to the leadership of citizens of Nordic countries in their use of e-government services, and this is not surprising, given the traditional leadership of these countries and their governments in the installation and adoption of telecommunications means, which we noted already (Kellerman 1999). Korea's high score may be related to its more recent predominance in the adoption of communications novelties, notably broadband (Kellerman 2006b). Interestingly, the governmental accent on the adoption of telecommunications services in France at the time has not been too successful as

*Table 7.3* Percentage of citizens in OECD (Organization for Economic Cooperation and Development) countries using the Internet to interact with public authorities, 2005 and 2010

| Country | 2005 | 2010 |
|---|---|---|
| Iceland | 55 | 75 |
| Denmark | 43 | 72 |
| Norway | 52 | 68 |
| Ireland | 18 | 67 |
| Sweden | 52 | 62 |
| Korea | 21 | 60 |
| Netherlands | 46 | 59 |
| Finland | 47 | 58 |
| Luxembourg | 46 | 55 |
| Mexico | 38 | 54 |
| Estonia | 31 | 48 |
| Canada | N/A | 46 |
| US | 23 | 41 |
| Slovenia | 19 | 40 |
| UK | 24 | 40 |
| Austria | 29 | 39 |
| Germany | 32 | 37 |
| France | 26 | 37 |
| Slovak Republic | 27 | 35 |
| Belgium | 18 | 32 |
| Spain | 25 | 32 |
| New Zealand | 32 | N/A |
| Hungary | 18 | 28 |
| Switzerland | 24 | N/A |
| Portugal | 14 | 23 |
| Poland | 13 | 21 |
| Japan | 18 | N/A |
| Czech Republic | 5 | 17 |
| Italy | 14 | 17 |
| Australia | 15 | 2011: 47 of those who used Internet |
| Greece | 7 | 13 |
| Turkey | 6 | 9 |

Data source: OECD 2012a; US 2010: OECD 2012b; Australia 2011: Australian Government Information Management Office 2011.

far as the use of e-government services, probably for the reasons mentioned above, and this might be true for the US, as well. The tremendous growth in the use of e-government services in Ireland from 18 percent in 2005 to 67 percent in 2010 may be attributed to the development there of an integrative system for all governmental authorities (Golden *et al.* 2003).

## Online banking

Banks, obviously, deal with money as a product to be bought, sold, exchanged and deposited. It is, thus, important to note, when assessing human performances in real and virtual action spaces that money has been the only material product so far which has been turned into information in the information age, and this transition emerged already before the introduction of individual online banking over the Internet. Thrift (1995: 27) noted in this regard: "nowadays money is essentially information." The first steps in the digital transmission of funds affected, as of the 1970s, bank offices only, with the introduction of electronic transfers of funds among banks, maintained through electronic funds transfer systems (EFTS) (see Malecki and Moriset 2008; Warf 2013). These inter-bank electronic transmissions of funds were followed, in the second phase of automatic banking, by the introduction of the automatic issuing of cash money to individual bank customers via Automatic Teller Machines (ATMs), as of the 1990s.

The introduction of the Internet in the mid-1990s opened up a third phase in automated banking, offering the possibility to establish online banking services, which have constituted virtual banking action spaces, and which have imitated real bank branches, in that they have facilitated the execution of bank operations, such as fund transfers, stock exchange activities, depositing and withdrawal from saving accounts, and so on. This latter phase had to await at the time the solution of security problems, side-by-side with proper domestic legislations and licensing. Finally, and more recently, banks have developed special versions of their websites as applications for smartphones, so that in the US in 2012, some 44 percent of smartphone owners used them also for banking activities (PewInternet 2012a). The introduction, though, of purely virtual, branch-less banks has failed several times, because specialized services, such as credit management, securities trading, life insurance and tax consulting, still require face-to-face meetings in real space branches, even for highly capable customers in the operation of online banking (Malecki and Moriset 2008). Still, however, there exist also some successful Internet banks, such as the Dutch RaboDirect.

The supply side of online banking is out there nowadays on the Web for most banks in the developed world in the form of sophisticated website systems, or virtual banking action spaces, and the question is if they are widely used by the demand side, namely bank customers. The question is simply: do Internet users prefer to access online banking and perform banking activities online? Worldwide, in 2012, some 59 percent of those connected to the Internet made use of the Web to "check bank account and other financial holdings" (Ipsos 2012a), but one can only assume that those who check bank accounts "passively" online would perform also "actively" bank activities online. For the US it was variously reported for 2011 that some 61 percent of adult Internet users did any banking online (PewInternet 2013a), or 68 percent did so in 2011–2012 (American Heritage Bank 2013), whereas in the EU in 2012 some 54 percent did so (Eurostat 2012a). Leading countries in the use of online banking are presented in Table 7.4.

The leadership of Sweden in the use of online banking is once again not surprising, as we noted already, and the percentage of Swedes adopting online banking is high, as compared to the adoption of other "e-" activities.

*Table 7.4* Leading countries in percentage Internet users performing online banking 2012

| Country | Percentage online banking |
| --- | --- |
| Sweden | 88 |
| France | 76 |
| Canada | 75 |
| Australia | 74 |
| Poland | 74 |
| South Africa | 74 |
| Belgium | 73 |
| UK | 71 |
| Korea | 70 |
| US | 70 |
| China | 61 |
| Germany | 61 |
| Turkey | 61 |
| Spain | 61 |
| India | 57 |
| Japan | 52 |
| Hungary | 51 |
| Italy | 51 |
| Russia | 48 |
| Indonesia | 45 |
| Argentina | 36 |
| Brazil | 32 |
| Saudi Arabia | 29 |
| Mexico | 23 |

Data sources: Ipsos 2012a.

Surprisingly, the percentage Americans and Koreans who use online banking is relatively modest, whereas the equivalent percentage of Chinese who do so is high. It seems that the use of virtual action space for banking involves not only a reasonable capability to use banking websites and to understand well the basics for financial activities, but it implies also customers' trust, mainly in the security measures provided by the banks. Online banking is, thus, more popular among the young, educated and more affluent customers (Ipsos 2012a, 2012b).

## Travel online

Travel online differs from the "e-" activities which we addressed so far, in that it does not constitute an end by itself, like online work, shopping and banking, but it rather constitutes preparatory activity for yet another activity which will take place in real space, namely the very travel away from home for business or for pleasure (see Kellerman 2012a). Browsing through websites which describe destination cities, hotels and attractions has become a routine action for many who prepare for out-of-town, whether domestic or international travel. The ability for individuals to reserve flights, hotels and rented cars over the Web has created a virtual action space styled as a travel agency. Still, however, even those who prefer to fully prepare their travel over the Web may need to consult a travel agent or an airline agent over the phone or face-to-face, in cases of complex travel itineraries. The preparations for travel are performed at times and spaces other than the flight itself, at home or in one's office, but such actions constitute indispensable components of the flight itself, because flights require reservations and tickets. Hence, the preparatory phases for travel and the travel itself jointly present a kind of disembedded hybridity between aerial and virtual activities and action spaces.

Similarly to the development of automated bank services, airline reservations have been digitized first for the airline industry and for travel agents through the introduction of electronic reservation systems (ERSs). Only later on, following the introduction of the Internet, such systems have become accessible to potential passengers through the Internet, either directly to specific airline systems, or to online travel agencies presenting routes and prices (see Malecki and Moriset 2008). Similar systems have been developed for railway and other terrestrial and maritime transportation systems, side-by-side with equivalent reservation systems for lodging and car rental services. The availability of online reservation systems for potential passengers has brought about mergers of real-world travel agencies and the closing down of others, so that less such retail offices are visible in urban landscapes.

"With the exception of finance, no economic sector more clearly demonstrates the pathbreaking power of Internet-based electronic platforms than the travel industry" (Malecki and Moriset 2008: 110). Thus, for the US in 2011 it was found that 65 percent of the adult users bought or made a reservation for travel online, and 52 percent took a virtual tour of a location online in 2012 (PewInternet 2013a). In Australia, in mid-2012, some 66 percent of adult Internet users used the Internet for travel purposes (IAB Australia 2012). In India in 2011 only 28 percent of Internet users booked travel online, whereas in China in 2010, some 62 percent of Internet users did so (Nielsen 2012). Korean authorities do not survey the specific use of the Internet for travel reservations made online, and they rather refer to shopping in general, so that in 2011 some 58 percent of Internet users in Korea performed shopping and selling over the Internet. Virtual touring in Korea is included in leisure activities, jointly with games, and this activity was performed in 2011 by some 88 percent of the Korean Internet users (KISA 2012).

For the EU in 2012 it was found that some 50 percent of the European Internet users made use of services that were related to travel online (Eurostat 2012a). The country specific data for EU member and candidate countries are presented in Table 7.5. The widest use of online travel services was made again in a Nordic country, Finland (with 69 percent of Internet users), and this rate might have also been the highest global rate for 2012, maybe jointly with the US and/or China. Finland was followed in 2012 by yet another small country, Luxembourg (with 65 percent of its Internet users), which on its part was followed by a big country, Germany (62 percent of Internet users). The Nordic leadership in the adoption and use of telecommunications services was noted previously. For Luxembourg the higher popularity of travel online might reflect frequent travels from this

*Table 7.5* Percentage Internet users in the EU obtaining travel and accommodation services online 2012

| Country | Percentage |
| --- | --- |
| Belgium | 49 |
| Bulgaria | 18 |
| Czech Republic | 58 |
| Denmark | 60 |
| Germany | 62 |
| Estonia | 28 |
| Ireland | 61 |
| Greece | 38 |
| Spain | 58 |
| France | 47 |
| Italy | 45 |
| Cyprus | 48 |
| Latvia | N/A |
| Lithuania | 21 |
| Luxembourg | 65 |
| Hungary | 25 |
| Malta | 46 |
| Netherlands | 55 |
| Austria | 51 |
| Poland | 20 |
| Portugal | 28 |
| Romania | 24 |
| Slovenia | 46 |
| Slovakia | 52 |
| Finland | 69 |
| Sweden | 58 |
| UK | 58 (2011) |
| Iceland | 54 |
| Norway | 59 |
| Croatia | 23 |
| Montenegro | 25 |
| Turkey | 19 |

Source: Eurostat (2012b).

small but most connected country, whereas in Germany it may relate to the large number of German outgoing tourists. East European countries seem to still lag behind in the adoption of travel online, as part and parcel of a more general slowness in the adoption of actions that can be performed over the Internet.

## E-learning

An additional virtual activity which has become available to individuals in the information age is e-learning, or as it is sometimes called, *distance learning*, offering academic (and other) studies through the Internet, in domestic institutes of higher learning or through cross-border education. Such studies amount to the attainment of codified knowledge, but the Internet further permits people to gain an enormously wide tacit knowledge, as well. Such tacit knowledge may be acquired by students through websites reached by Web browsing, or through virtual consultations with remotely located colleagues, possibly enriching their traditional frontal study. Codified knowledge, on the other hand, may be gained online either through study in single formal courses or through full academic degree studies. The acquisition of tacit knowledge is informal, by its very nature, it may range in volume from time to time, and it is similar to other forms of informal personal communications and networking among individuals. Formal degree studies, or purchases of codified knowledge, might be viewed as being similar to e-commerce, in that a product, knowledge, can be bought online. There are, however, some differences between the sale of products and services (such as airline tickets) over the Internet, on the one hand, and the provision of formal codified knowledge through the Web, on the other, and these differences may eventually bring about differing levels in the provision of online formal education, as compared to purchases over the Internet of products and services. Some of these differences are highlighted in the following paragraphs.

From the supply side, language is a much more crucial element in e-learning than in e-commerce, by the very nature of the learning process. Standards of quality are another dimension of formal study differing from standards of products and services. Some internationally recognized quality standards for products and services (e.g., ISO9000) have been widely adopted, but domestic or cross-border academic degrees have to be recognized by national higher education councils, as well as by potential employers, without an availability of widely recognized standards for academic courses and degrees. Efforts have been made in recent years at the international level of high-education for the establishment of some standardization of academic study under the circumstances of an opening and globalizing world, such as the EU Bologna agreements, as well as global GATS (General Agreements on Trade in Services) agreements (Knight 2006). In other cases, the prestige of globally leading universities may bring about domestic recognition, though some professions may require in any case some domestic adjustments (such as law and medicine).

From the demand side of potential students, the purchase of an online degree program implies a prolonged purchase process spanning several years, as compared to a few minutes when buying other products and services online. E-learning further requires a much more extensive investment by the customer, involving financial costs of tuition, extensive time allocation, and intellectual efforts, notably when studying in virtual and rather individual action spaces. These latter special intellectual efforts may make online studies look inferior to face-to-face ones, from the perspectives of students and universities and colleges alike. As far as product durability is concerned, an academic degree constitutes a lifelong product, as compared to the disposability or limited lifespan of commercial products and services.

For all these reasons, and additional ones, online formal academic education has turned into a widely-debated issue, coupled with its being still in relatively modest use, even as compared to e-commerce. This modest use of e-learning is further related to the fact that distance learning is relevant only to potential students, whereas shopping online and e-banking may be attractive and adopted by society at large in developed countries (for debates see, e.g., Breton and Lambert 2003, and for a review see Huh 2006). It has turned out most difficult to find internationally comparative data on distance learning, notably most recent ones, since e-learning refers jointly to both academic and to non-academic studies, as well as to both full and part-time study, so that its internationalization is still in its infancy.

The *eUSER Population Survey 2005* carried out by the EU (2006) in ten member countries sheds some light on trends in e-learning in Europe several years ago, but the survey did not inquire about online full-degree studies. The survey yielded low results for the more modest option of online studies of at least one course (Table 7.6). In terms of the percentage of e-learners out of the total adult learners (and not out of the total population), the results ranged from the rather low 3.3 percent in Germany, which ranked much higher in domestic e-commerce (see Table 7.1), to the leading UK (10.7 percent) and Ireland (10.0 percent). The relatively higher values for the UK and Ireland may be related to the wider availability of study opportunities in English, as well as to the impact of the Open University in Britain. The need to overcome traditions, habits and conventions by potential students, and the hesitation of many universities and colleges to fully teach over the Internet, require careful attention in the tracing of possible future developments of study opportunities over the Internet.

In the US, back in 2000–2001, some 56 percent of the degree-granting institutions offered distance courses, but not necessarily full-degree studies online. These offerings ranged widely among university and college types. Hence, just 16 percent of the private two-year colleges offered such courses, as compared to 90 percent of the public two-year colleges. By the same token, merely 40 percent of the private four-year institutions included online courses, as compared to 89

*Table 7.6* Online course learners as percentage of adult learners 2005

| Country | Percentage |
|---|---|
| UK | 10.7 |
| Ireland | 10.0 |
| Denmark | 7.5 |
| Poland | 6.5 |
| France | 6.0 |
| Slovenia | 5.2 |
| Czech Republic | 4.8 |
| Italy | 4.7 |
| Hungary | 3.9 |
| Germany | 3.3 |
| Overall sample | 6.4 |

Data source: EU (2006).

percent of the public four-year institutions (US Bureau of the Census 2006). The major differences between the public and private sectors of universities and colleges in the US reflected academic differences regarding e-learning in general and concerning their responsibility towards their potential clientele, with the two-year colleges more inclined to foster higher education in peripheral regions and to cater for working students who need more flexible learning times. These data, reported by the US Bureau of the Census, do not permit to assess the percentage of students who took at least one distance course, but the data on the offerings permit to assume that this form of study was, at the time, more popular in peripherally located and low-income segments of the population, so that the Internet provided a learning opportunity for students who otherwise would not have been able to study for an academic degree.

Seven years later, in 2007–2008, the latest year for which official US data have been available, some data on the demand side for academic e-learning were revealed. Thus, 20 percent of the undergraduate students by then took at least one e-learning course, rising from 16 percent in 2003–2004, but the percentage of students who studied for their degree entirely through e-learning declined over this time period from 5 percent to 4 percent! At the graduate level in 2007–2008, the percentage of students taking remote courses was slightly higher, reaching 22 percent, but the percentage of those who studied entirely through e-learning was significantly higher, reaching 9 percent (Aud *et al.* 2011). For the year 2010, a higher demand for online courses, as compared with face-to-face ones, was reported for the United States (Hanover Research 2011). Apparently, though, basic academic education, as compared to advanced education, has presumably been perceived as mostly needing at least some face-to-face contact in real space (Aud *et al.* 2011).

There are universities, such as MIT, which have posted on the Internet the lectures of their entire faculty, for the purpose of self non-degree study, side-by-side with the numerous universities that offer full or partial formal e-learning services. However, the adoption of this option by students has been partial, even in the United States, which pioneered at the time in distance learning. In the 2010s a new trend has emerged in distance learning, namely the offering of Massive Open Online Courses (MOOCs), or full online interactive courses that can be used freely or for tuition on a course-by-course basis, rather than as part of a full degree program. Thus, in 2011, two academic entrepreneurs, Daphne Koller and Andrew Ng, established a global academic online study service, Coursera (coursera 2013), which began by offering courses of leading American universities on a full online basis, including interactions among the students themselves, as well as offering paid-for credit for these courses. By the end of 2013, over 100 universities and institutions worldwide joined the service. However, a student still cannot yet get recognition by a degree granting university for the completion of all the requirements for an academic degree on the basis of the collection of courses studied eclectically in numerous universities online.

## E-health

The use of the Internet by individuals for health-related matters has received numerous names, both general and specific, sometimes interchangeably. *Telemedicine* is the oldest term, dating back to pre-Internet times, when telegraphy and telephony were the major telecommunications media, and it was, thus, defined as: "the use of telecommunication technology for transfer of medical data from one site to another" (Pal *et al.* 2002). Such transmissions were, obviously, not necessarily made by layman individuals, and at the pre-Internet times such transmissions of medical information were mostly the responsibility of medical practitioners. Currently, telemedicine has turned into a more specific term referring to distance medical treatments (Smith 2004), and, thus, defined as "the use of advanced communication technologies, within the context of clinical health, that deliver care across considerable physical distance" (Breen and Matusitz 2010). The term *e-health* was considered, at times, as related to the provision of health information mostly as a commercial service (Pal *et al.* 2002), and later on more generally as the supply and availability of medical information online, including information on diseases medications, etc. (Breen and Matusitz 2010). *Telehealth* was suggested as an even broader term, including, for instance, distance learning for health professionals (Smith 2004). Others, referring mainly to the online search by individuals for medical information, suggested the term *health online* (PewInternet 2013d).

We refer here to the term e-health as a wide spectrum of health-related online services, which include the use of the Internet by individuals for three main health-related activities: management, consulting and treatment. First, the *management* of one's health includes activities such as appointment fixing with doctors, viewing the findings of laboratory and other medical tests, and ordering and receiving medical prescriptions. Second, medical *consulting*, or health online, which involves, above all, people's search for Web information on health issues, such as diseases, medications, etc. Third, medical *treatment*, or telemedicine, performed at a distance, through the Internet, such as teleradiology.

As compared with other virtual action spaces e-health is not fully automatic, such as e-banking, travel online and e-government. For many functions it involves people, such as expert physicians or other medical workers, performing treatments or sending prescriptions, on the other end of the line, either online with the patient or offline. Still, however, health management and consulting activities are similar in most of their functions to other e-activities, but telemedicine is different in that it requires the development of specialized equipment, software and specialists for its operation, more than is the case for shopping, or for government and travel services. Online social networking, which we will discuss in detail in the following chapter, may treat health issues, either in dedicated networks or in more general ones, thus complementing or replacing searches for health information on the Web (Griffiths *et al.* 2012).

The space programs operated by NASA required the development of technologies and specialists for operation under extreme distance and environmental conditions, and these developments were adapted for on-earth more routine distance uses (Matusitz and Breen 2007). Telemedicine technologies may, therefore, be helpful for treatment provision in extreme environments, such as sea oil drilling and pumping, and remote army bases, but they are of much wider applicability for rural and peripheral areas, notably those with extreme climate conditions, mainly in developed countries, and for developing countries in general (see Pal *et al.* 2002). However, in most cases the patients themselves do not operate telemedicine, and nurses or technicians are needed at the patient's end, side-by-side with physicians and specialists at the remote hospital or clinic end. As such, telemedicine does not amount to individuals acting directly through or in virtual space.

Several negative implications were outlined by Matusitz and Breen (2007: 97) for the use of telemedicine, and of e-health in general:

(1) lack of real-time interaction between the patient and provider, (2) the possible lack of reliability or accuracy of information provided by e-health resources, (3) the public health concern over consumer use of the Internet to self-diagnose conditions that may be life-threatening, (4) the inaccessibility to e-health by a certain number of disadvantaged and isolated groups, and (5) the deep intercultural differences among the patient and the healthcare provider that e-health services have not improved.

There are further several problems involved in the very operation of telemedicine such as insuring the practitioners acting in distance from the patients; recognition of health problems discovered at distance for payment or reimbursement by health insurance providers; frequent lack of expertise in the operation of technology and equipment in remote areas (Breen and Matusitz 2010); assuring the confidentiality of medical information transmitted over the Internet; and cross-border licensing of medical practitioners (Pal *et al.* 2002).

A wide survey of telemedicine services conducted by the World Health Organization in 114 countries in 2009 (WHO 2010) has shown that the most popular telemedicine medical service/treatment was teleradiology, available in 62 percent of the countries in established, pilot, or informal modes, followed by telepathology (in 41 percent of the countries in all modes), teledermatology (in 38 percent of the countries in all modes) and telepsychiatry (in 24 percent of the countries in all modes). However, these services were widely established in high-income countries and much less so in middle and low-income countries. Geographically, African and Eastern Mediterranean countries were the lowest ones in the availability of established telemedicine services and these countries simultaneously had a higher proportion of informal telemedicine services. Thus, the very potential ability of telemedicine to provide medical services in lower income countries has only partially been materialized, so that an international digital gap in this area still widely existed in 2009.

As far as e-health is concerned, with the connotation of Internet users searching personally for medical information on the Web, in 2013 some 72 percent of

American adult Internet users did so (PewInternet 2013d), and this is a high percentage, as compared to other uses of the Internet, being similar to the share of
American Internet users who shopped online. However, shopping online, as well
as in other online action spaces, involve some activity performed by Internet
users whereas e-health in terms of information search and gathering, seems to be
similar to other mere information searches related to prudential curiosity (see
Chapter 5). The Web search for medical information has been expanded with the
wide adoption of smartphones. Thus, in 2012, one in three American mobile
phone owners used their phones to look for health information, as compared to
only 17 percent who did so in 2010. However, among American smartphone
owners, 52 percent did so in 2012, as compared to merely 6 percent of non-
smartphone mobile phone owners in that year (PewInternet 2012b). Unfortunately, no comparative international data on e-health information searches could
be found.

## Comparative features and trends

We have reviewed so far some seven activities that are performed already fully
or partially online by many Internet users. We further noted before several differences among some of these activities. We will examine now all these activities together from a comparative perspective, using the observations we made in
the previous sections. Daily human activities differ from each other in their
nature and significance in people's lives. Thus, work extends for most adults
over at least one-third of each day during the week, and it is by far a leading
human activity with major implications for people's present and future. This
crucial significance and role of work for individuals may have led to a continued
global hesitation by workers to divert their physical locations while at work,
fully or partially, from their offices to their homes, although the work activities
of many workers is performed in virtual space, whether out of a computer
located in office or from the same one or a similar one located at home.

As we noted already, the purchase of an online degree program is somehow
similar to work in its being a rather prolonged purchase process, spanning several
years, as compared to a few minutes required for the purchase of other products
and services online. Online study further requires a much more extensive investment by the customer–student, in terms of its financial costs, the extensive investment of time, and the required intellectual efforts, notably when studying in virtual
and rather individual action spaces, as compared with being part of a physically
present group assembled together in a real classroom. This latter intellectual effort
involved in online study may make it considered as inferior to face-to-face ones,
from the perspectives of both students and universities and colleges. Another difference between learning and the purchase of goods and services is their durability,
which for an academic degree is without expiration, as compared to the disposability or limited lifespan of commercial products and services.

Shopping online is a more widely adopted activity than learning online, but it
is so in a more partial way, because a preference for shopping in real space still

prevails for certain products, and shopping online maybe split between browsing and shopping, with possibly just one of these phases taking place in virtual space. As compared to the wide choice of merchants and merchandise typical of shopping online, e-government is a rather compulsory and uncompetitive service, forcing citizens to receive governmental services, such as the payment of taxes and fees, submission of applications for permits, and the receipt of documentation, from the specific governments of their local, regional and national locations and their respective virtual action spaces.

Online travel activities differ from other "e-" activities in that these virtual activities do not constitute an end in themselves, like work, shopping and banking, but they rather constitute preparations for an activity which will eventually take place in real space, namely travel away from home, either for business or for pleasure.

E-health is not a fully automatic action space like e-banking, travel online and e-government. For many functions it involves people, such as expert physicians or other medical workers, performing treatments or sending prescriptions, on the other end of the line, either online with the patient or offline. Still, however, health management and consulting activities are similar in most of their functions to other e-activities, though telemedicine is different in that it requires the development of specialized equipment, software and specialists for its operation, more than the case for shopping, or for government and travel services. Furthermore, in most cases nurses or technicians perform the treatment at the patient's end. The mere search for medical information does not involve any action by Internet users, as is the case with shopping online as well as in other online action spaces. As such, e-health in terms of information search and gathering, seems to be similar to other mere information searches related to prudential curiosity.

Close to the middle of the second decade of this millennium it seems that Internet users have shifted part of their daily activities in numerous major spheres of life from real to virtual action spaces, though the percentages of users who make use of virtual action spaces still differ widely from activity to activity and from country to country. The leading country percentages of Internet users who perform daily activities online are already quite high for most of the activities which we reviewed in this chapter, activities which require instant and short-spanned action, and activities which are mainly commercial ones, including shopping, banking, leisure bookings and the non-commercial interaction with governments (Table 7.7). However, the percentages of use for activities which require prolonged diversion of action from a real to a virtual action space are still much more modest. This is mostly so with regard to work and for learning. As far as these activities are concerned, people still prefer, at least partially, face-to-face contacts.

The leading countries in the performance of activities in virtual action spaces are the US, which has been the major inventing and introducing country for most new information technologies, and the Nordic countries which have traditionally been the pioneering countries in the massive adoption of new devices and uses.

*Table 7.7* Online major commercial activity spaces 2010–2012 by leading percentage
         actors and countries

| Activity | Year | Leading percentage users | Country |
|---|---|---|---|
| Almost full home-based work | 2010 | 4.9 | UK |
| Online shopping | 2011 | 71 | US |
| E-government | 2010 | 75 | Iceland |
| Online banking | 2012 | 88 | Sweden |
| Travel online | 2012 | 69 | Finland |

Data sources: See text and Tables 7.1–7.5.

These two traditional regions in the development and adoption of IT newness
have been coupled with two Asian countries which seem to lead in the use of
virtual action spaces via mobile broadband. Korea was the pioneering country in
this regard, and it has recently been joined by China.

## Conclusion

The review of the several activities that are contemporarily performed by individuals also online points to complementarity between real and virtual action
spaces rather than to full substitution of the real by the virtual. Thus, for each of
the reviewed activities there was shown some advantage in doing business
online, but at the same time there still exists some or much need for action in
real space, as well. This is even more striking when taking into account the
growing use of mobile access to the Internet, enhanced by an ease of use offered
through the smartphone and tablet versions of browsers and websites (through
applications). These have made it possible for users to consume services without
limitations of time and location in their routine daily life, thus amplifying even
further the blurring separation of the temporal and locational dimensions of daily
activities. However, gaining experience in the use of the growingly sophisticated
smartphone applications, may potentially, at least, change the balance between
services offered through facilities located in real space within easy physical
reach by users, on the one hand, and virtual services provided through the Internet, or services which have been located anywhere in real space and can now be
reached through the Internet, on the other. Such possible changes would, of
course, mostly be in favor of the virtual at the expense of the real and close by.

   The relatively veteran virtual action space of work has turned out to still be
the most modest one as far as its full adoption is concerned. Trends in both the
US and the EU point to constant growth in the number and percentage of those
working from home, including those who work only or primarily from home, but
this pattern of location for work is still less-preferred, thus still consisting merely
single-figure percentages of the workforce. In other words, only few people in
countries that lead in the personal adoption of Internet connectivity are ready to
use a virtual action space exclusively for their work, and the vast majority of
workers still prefer a mix of physical and real action spaces for their work.

Face-to-face work and work supervision are still highly valued by employees and employers, respectively.

The data for online shopping presented percentages of the population doing some but not all of their shopping online. The overall picture of the seemingly modest percentages of population performing online shopping, mainly in the EU, seems to suggest that the geographical location of customers and their preference for traditional stores located in real space are still of significance, and this preference might still continue into the future. This tendency for shopping in real action space, pertaining to shopping by individuals, is there because of shoppers' attitudes. Habits, trust and uncertainties turn out to constitute main obstacles in the way for the adoption of online shopping, side-by-side with shoppers' preference for touching and trying of merchandise, coupled with their perception of shopping in real space as being also a kind of entertainment.

E-government has become widely offered and adopted in developed countries, but it awaits even further growth with expected future declines in the prices of IT hardware and services, coupled with a growing ease of use of e-government, thus making it more affordable to the elderly and the poor. Lack of trust by citizens in governmental e-services might also be an obstacle for a wide e-government adoption, and the lack of competition in the offering of e-government services might further slow down the pace of development of otherwise relatively easy to use online services.

Another, and rather commercial, activity which requires much trust is banking. It seems that the use of virtual action space for banking services involves not only a reasonable capability of bank account owners to use banking websites and to understand well the basics for financial activities, but it implies also customers' trust in banking services, and much more so than the trust required for real space banking services, mainly because of the more extensive security measures required for online banking transactions.

The travel industry facilitates the purchase of travel services online, and this involves the application of online financial transactions similarly to those used for shopping online. However, the information which customers would consult online prior to the placing of travel reservations and the financial transactions for their costs is much more extensive than the one required for the shopping of material products. Such pre-order information is required on potential places and attractions for pleasure travel, as well as on potential service suppliers, such as hotels, and this in addition to the rather complex information that is required on the travel itself, mainly on flight schedules and prices. Since the major feature of the Internet as an action space is its being an information space, travel online has developed fast and may develop even further in the future, notably with regard to the supply of touristic information.

Distance learning might well be the most complex action space in its attempt to sell knowledge, as compared to the sale and provision of products, services and information, pursued by other virtual action spaces. Distance learning is provided not only for the provision of academic knowledge, but for the provision of professional one as well. It seems that here too the preferred mode for both students and institutions of higher learning is some mix of face-to-face courses and video distance ones, whether pre-recorded or live ones.

E-health includes a wide variety of virtual activities consisting of management, consulting and treatment of one's health. The management of people's health life, as well as their consulting of health-related Web information, has become routine activities in developed countries. The management component of e-health consists of an action space, similarly to other daily "e-" activities, whereas Web searches for health-related information are part of the wider tendency to satiate one's prudential curiosity through the Web. Telemedicine, or distance medical treatments, are relevant mostly to rural and peripheral areas in developed countries, and to developing countries at large. However, despite the availability of telemedicine technologies, there still exists a digital gap in this regard between high-income countries and others.

# 8  Social networking

Technology proposes itself as the architect of our intimacies. These days, it suggests substitutions that put the real on the run.

(Turkle 2011: 1)

Web 2.0, or virtual social networking, will constitute the focus of the discussions in this chapter. Whereas the previous chapters focused on personal and economic needs of individuals, our attention here will move to people's social arena, namely meetings among people, public ones as well as more intimate ones. Aristotle recognized in his well-known *Politics* (originally written in 350 BC) that social contact constitutes a basic human need when he stated that "Man is by nature a social animal."

The structure of this chapter will lead us through several discussions. First, social networking will be outlined in general, and this will be followed by a discussion on the emergence of online social networking, side-by-side with an exposition of the patterns and rules of this latter form of networking. Special emphasis will be put on the widening of human relations through online networking and the social significances of these wider relations. Second, the several phases in the development of Internet social networking will be explored, and this exposure will be followed by an examination of dimensions of virtual versus real social action spaces. Some general features of real and virtual social action spaces as two seemingly distinct action spaces, but still interrelated with each other, will be discussed, and some internationally comparative data on the adoption of Facebook, as the leading online social networking platform, will be presented and interpreted.

## The nature and significances of social networking

Rainie and Wellman (2012: 21) portrayed contemporary online social networking from a societal perspective, and they identified the following general nature of social relations:

Society is not the sum of individuals or of two-person ties. Rather, everyone is embedded in structures of relationships that provide opportunities,

constraints, coalitions, and work-arounds. Nor is society built out of soli-
dary, tightly bounded groups – like stacked series of building blocks.
Rather, it is made out of a tangle of networked individuals who operate in
specialized, fragmented, sparsely interconnected, and permeable networks.

Online networks, like those in real space, are supposed, therefore, to possess
numerous qualities for their participants, and they may facilitate the emergence
of a social action space for varied activities. Online social networking possesses
yet another special feature which is its very use of information technology. Thus,
social networking websites were defined by Boyd and Ellison (2007, see also
Davis 2010) as:

> web-based services that allow individuals to 1) construct a public or semi-
> public profile within a bounded system, 2) articulate a list of other users
> with whom they share connection, and 3) view and traverse their list of con-
> nections and those made by others within the system.

Thus, the Internet permits individuals to maintain wide geographical and social
action spaces, and it further facilitates the very management of wide social contacts, as well as the management of the contents shared among these and other
contacts. These traits of online social networks stem from the very constitution
of the Internet as an informational action space, and such traits are, therefore,
normally not used for real space networks.

Social networking has been viewed as a leading component of the wider trend
of online networking, which Benkler (2006: 3) termed the "networked information economy." This new networked information economy has displaced the
industrial information economy, and has been characterized by "decentralized
individual action – specifically, new and important cooperative and coordinate
action carried out through radically distributed, nonmarket mechanisms that do
not depend on proprietary strategies" (Benkler 2006: 3). Tapscott and Williams
(2010: 30) identified four principles for this new networked informational
economy: openness; peering; sharing and global action, and these as compared
to "the hierarchical, closed, secretive, and insular multinationals that dominated
the previous century." Social networking, as part and parcel of the more general
networked information economy, may be viewed, therefore, as individual
actions, coordinated with other people's individual actions, and aided by the distributed mechanisms that are provided through information technology over
proper platforms. These social and technological elements may jointly bring
about social ties of numerous types and forms, and by the very nature of the networked information economy these networks permit more democratic cultural
production.

At the heart of the new networked economy stand, therefore, the networked
individuals, namely people who are "networked as individuals, rather than
embedded in groups" of whichever type, such as families, work units, neighborhoods, or social groups (Rainie and Wellman 2012: 6). Furthermore, these

networked individuals present a new sense of personal, albeit connected, autonomy, expressed also through increased personal mobilities (Rainie and Wellman 2012; Kellerman 2012a). Networked individuals may interact with several people at a time, while simultaneously being also engaged in other activities. However, online social networking is demanding, so that "networked individualism is both socially liberating and socially taxing" (Rainie and Wellman 2012: 9). It is liberating mainly through the immense widening of social ties, but it is also taxing in the time and effort it requires, and potentially, at least, also in the weakening of ties in real space which it may bring about.

The very nature of the social relations developed online through social networks, as well as their potential impact on social relations in real space, has been widely debated in recent years. This debate has taken place in parallel with the massive adoption of social networking by Internet users worldwide, so that societies worldwide are still in search of balances between real and virtual social action spaces and social relations in them. Furthermore, since Web 2.0 and the major network platforms based in it are still new, it might be too early to firmly and decisively describe and assess the new world of social relations, activated and existing simultaneously in both real and virtual space. The relations between social real and virtual action spaces seem to be still in their evolutionary phase, and are, thus, still unsettled. Differences and nuances among scholars in the assessment of social relations in this early period of online social networking may stem also from the rather wide variety of the disciplinary affiliations of the scholars involved in the assessment of the new world of online social networking. As far as the researchers whose works will be mentioned in the following paragraph, the following disciplines are represented: Law (Benkler); Clinical Psychology (Turkle); Sociology (Wellman); Political Science (Rainie) and Geography (Warf).

For Benkler (2006: 371) the Internet "simply offers more degrees of freedom for each of us to design our own communications space than were available in the past." This wider choice of communications media has implied, á la Benkler (2006: 15), that the Internet has been used at the expense of television, and hence strengthening ties with family and intimate friends, whether located near or far, side-by-side with the establishment of wider, and more diversified, but also weaker new ties (see also Rainie and Wellman 2012: 127). Turkle (2011: 14) seems to be more doubtful in this regard: "these days, whether you are online or not, it is easy for people to end up unsure if they are closer together or further apart," but for Rainie and Wellman (2012: 127) it is evident that "the more Internet contact, the more in person and phone contact," among both family members and friends. Another potential possibility for close social relations, notably as far as their fostering in real and virtual spaces is concerned, is that online networking may facilitate the development of stronger ties among individuals, as long as there exist parallel social ties among these individuals in real space (Warf 2013). Such ties may permit the emergence and the development of *emotive trust* among communicating parties (Ettlinger 2003). However, given the richness of contact associated with face-to-face meetings by their very nature, real space "still retains a vital role in contemporary economic and social life" (Warf 2013: 147).

Social networking has contributed to significant changes in people's personal lives, as well as to societal–political events. Personally, and relating to romantic relations, Ben-Ze'ev (2004) presented the evolution of online romantic and sexual relations all the way from initial electronically-written contacts through email, real-time exchanges, blogs and SMSs, followed by video conversations, to cybersex activities. Socially, online communications channels provide for free expression and exchanges even in countries with limited democracy, with a special significance for women for whom the Internet may sometimes constitute an alternative to other, more restricted, channels for reaching out. At the societal level, online social networking systems have served as major tools in the provoking and the sustaining of political unrest and revolutions, such as during the 2011 Arab Spring events in Middle Eastern and North African countries, following similar earlier events in Russia and Iran.

## Online social networking systems

Castells (2000) identified and developed the notion of *network society*, based mainly on the "space of flows" consisting of business ties among cities globally. However, in parallel to this network society of business which includes also elite segments of society, there has gradually emerged a more popular network society, which opened its gates, mainly through Web 2.0 applications, to all interested individuals who have been permitted to do so by their governments and by their cultural values (see Chapter 9), and this latter widely spread process has brought about the more general networked information economy (Benkler 2006).

Social networking nested within the email and Gopher systems even before the inception of the Internet and its declaration as a wide and open access system in the mid 1990s. Thus, some global networks developed initially around a physical location, e.g., the San Francisco-based *WELL* network (see Rheingold 1993), whereas others, such as *MOOs*, were organized around metaphorical cities, thus ordering centrality and agglomeration in the volume and intensity of communications to specific "rooms," "buildings" or "neighborhoods" (see Schrag 1994).

A second generation of social networking, becoming popular as of the mid-1990s, consisted of online exchanges, via systems such as the two popular ones at the time, ICQ (I Seek You) and MSN (Microsoft Network), followed by blogs which were instituted as of 2002 (Herring *et al.* 2005), but were basically initiated in different forms much earlier (Gopal 2007). Blogs constituted, at the time, also part of the first generation of globally wide and free self-publications of personal materials of all types by individuals over the Internet, spreading widely through blog interlinking, within what was termed the "blogosphere." In this feature of messages spreading over the virtual action space, as well as in their facilitation of the joint production and usage of knowledge and information, blogs differ from personal websites, which we discussed in Chapter 6. Personal websites normally are not interlinked, and thus are more "passive" in their virtual mobility or spread, and they rather "await" to be accessed by interested users (Bruns 2008; Jones *et al.* 2010; Warf 2013).

It was estimated that about 100 million blogs worldwide existed by October 2005, including a high percentage of spam (*The Blog Herald* 2005), and their number grew to some 173 million in 2011 (Warf 2013). International differences in the country distribution of blogs in 2005 existed within both the developed and the developing worlds. The US dominated the blogosphere scene with some 30–50 million blogs, while neighboring Canada had only approximately 700,000 ones. Similar gaps were witnessed in Europe as well. Thus, France had at that time some 3.5 million blogs, whereas Germany had only 300,000. By the same token, Spain and Poland had 1.5 million blogs each, whereas in Italy there were only 250,000 ones. Leading industrializing countries of the age presented similar trends of high international differences, with China having 6 million blogs, while India had only 100,000 (*The Blog Herald* 2005). Local cultures and tendencies have played, therefore, a leading role in the adoption rate of blogs, something that has become even more striking with the later adoption of Facebook, the data for which we will present in a following section of this chapter.

The social networking dimension of the Internet has become extremely popular in a third phase of social networking commencing with the emergence of Web 2.0, the virtual framework which has flourished as of the late 2000s with a focus on social networking. Web 2.0 has hosted since then several swiftly adopted networks or platforms for online social networking, led mainly by Facebook, Twitter, MySpace, LinkedIn and Second Life. The last one, Second Life, differs from the other online networking platforms in that users deliberately create avatars that communicate with each other, whereas in the other networks this is only an option, as we noted already in Chapter 6.

Facebook has become the major competitor for Google on the dominance of the Web in terms of the services offered and the number of customers drawn to use them, with Google specializing mostly in Web 1.0 applications, notably search engines, and Facebook specializing in Web 2.0 applications, and more specifically in social networking. Facebook was established in 2004, and it offers subscribers several services, including mainly the presentation of personal profiles, side-by-side with the presentation of commercial pages and websites; the online publication of personal materials, distributed to specific friends subscribed to the Facebook system; and the rating of other subscribers' writings. Back in 2009, some 200 million people or 13 percent of Internet users worldwide used Facebook actively, and one half of them did so at least once a day (Scherr Technology 2009). As we noted already, just four years later, in 2013, it was estimated that about 43 percent of global Internet users (one billion out of 2.3 billion) were Facebook subscribers (Checkfacebook 2013), as compared to 100 million Twitter active users! On the average, each Facebook subscriber in 2012 had 130 friends, and she/he liked 80 pages (digitalbuzz blog 2013).

There are probably a significant number of Internet users who are subscribed to more than one social networking system/platform (see Rainie and Wellman 2012: 12). It was further estimated for the US that some two-thirds of its adults and some three-quarters of its adolescents have created online materials (Rainie and Wellman 2012: 14). We may, thus, quite safely assume that by 2013 some three-quarters of the Internet users worldwide were engaged in social networking over the net. Thus, the barrier-free, constraint-free (in most countries) and cost-free action space for virtual social networking has exhibited a tremendously fast adoption rate. However, the very use of social networking over the Internet does not automatically imply globally stretched networks created and attended by all of its subscribers. Facebook has become, for example, a framework for virtual interaction among school kids whose location may not stretch beyond a single neighborhood (see Rainie and Wellman 2012: 130–1).

## Real versus virtual social action spaces

We noticed in a previous section the still indecisive views among students of social networking regarding the possible impacts of online social networking on social networking and social relations in real space. In this section we would like to elaborate more generally on real and virtual social action spaces as two seemingly distinct action spaces, but which are still interrelated. We will, thus, briefly elaborate in the following paragraphs on the following aspects: loacational, temporal and distance dimensions of the two social action spaces; the transforming balance between face-to-face and virtual social relations with the adoption of email, followed by the fast adoption of online social networking; the silence and noise dimensions of the two forms of social interaction; the possible emergence of placelessness among individuals heavily involved in online social networking; and, finally, the special case of networks aimed at creativity.

Wellman (2001, see also Rainie and Wellman 2012) outlined several phases in the process of change in the conduct of social relations following the adoption of communications technologies. The Internet and wireless communications were viewed as expressing a new, and rather third, phase in social communications and networking. The first phase of social relations was suggested to constitute the traditional and non-technological communications of people, when walking for visits with each other. It was termed by Wellman (2001) as *door-to-door* communications, constituting obviously face-to-face relations. These visits typified social relations within traditional, physical-place bounded, communities and action spaces. This type of communications required *synchronous presence* of the communicating parties in real space (Yu and Shaw 2008). The automobile and the telephone have permitted the development of a second phase of social relations and networking, namely *place-to-place* ones, offering some flexibility in the location of people's social relations, and thus replacing some of the local door-to-door relations. Place-to-place communications consisted, therefore, of both face-to-face ones in real action space using cars, alongside virtual ones over the telephone, or in a kind of virtual action space, the latter being termed as *synchronous telepresence* (Yu and Shaw 2008).

The Internet has enhanced place-to-place networks through its provision of the option for continuous communications. However, the most significant contribution of the Internet was that placeless wireless communications have implied the emergence of a third phase in social networking, namely that of *person-to-person* communications, with the communicating parties optionally detached from their household locations and from the communications infrastructure of their homes. This person-to-person communications has been further enhanced with the widespread mobile broadband communications. As compared to the telephone, Internet communication is both locationally and temporally flexible, since it does not require the synchronous attendance of the communicating parties. It was, thus, termed *asynchronous telepresence* (Yu and Shaw 2008).

Traditional interpersonal communication has involved strong spatial elements, through face-to-face meetings in real space, side-by-side with postal communications that carried clear geographical connotations, notably through national stamps, the use of street addresses on the envelopes, and the mentioning of the location of writing on top of letters. The elements of real space in electronic communications have obviously been much weaker, but they have not vanished. As we noted already, electronic communications media mostly do not constitute stand alone forms of communications, since they are anchored within spatial facilities and devices, and more importantly, repeated virtual communications among parties may frequently lead to face-to-face meetings of the parties (see Boden and Molotch 1994; Urry 2002; Tillema *et al.* 2010). Thus, human relations may be viewed in many cases as stratified through the adoption of varied communications technologies. Some specific exchanges, such as romantic ones or business-oriented ones may begin with written communications, and if fruitful may move on to vocal contacts over the telephone, and only if this phase proves satisfactory then face-to-face contact is called for.

The decline in the significance of location for social relations has been coupled with viewing the second basic spatial element, distance, as weakening too in our daily lives (Cairncross 1997). However, as was shown by Mok *et al.* (2010), in their study of Toronto, distance was still significant for human social relations, just before the massive adoption of online social networking. Comparing communications performances in the same Canadian urban setting in the years 1978 and 2005, they showed that distance has been still significant for communications. The introduction of email in the 1990s has brought about an increase in communications activity at large, with emailing preferred over face-to-face and telephone contacts. However, though emailing by its very nature is insensitive to distance, and thus may potentially lead to long distance communications, Mok *et al.* (2010) were able to show that the significance of distance for face-to-face and telephone contacts has remained unchanged between 1978 and 2005.

The fast and wide adoption of Facebook as of 2005 may have potentially changed, though, these long standing trends. Turkle (2011: 13) noted in this regard, that online connections were conceived initially as substitutes for face-to-face (or door-to-door) meetings and telephone calls (or place-to-place relations), but they have soon turned into the connection mode of choice, aiding people with overworked and overscheduled life, side-by-side with their facilitation of loneliness overcoming. By the same token, Rainie and Wellman (2012: 13, 130–1) believe that online social relations with geographically close friends are still the leading ones, thus turning online networks into the new neighborhood. However, there is a possibility, of which Turkle (2011: 154) warns, that individuals removing themselves from physical social life because of online-networked relations, may make such Internet users "become less willing to get out there and take a chance."

Another dimension of social interaction in real and virtual action spaces is noise/quietness. Social interaction that takes place in real space facilities and action spaces, such as bars, cafés, etc., is by its very nature noisy, at least in some way. On the other hand, however, social interaction through the Internet may imply a choice between either quiet or silent contexts, as we have seen in Chapter 1. Quiet contexts apply to audio conversations between two people located within a silent context and using audio telephone calls (VoIP). Silent social interaction is the case when exchanges are made only in writing, whether through chat platforms (e.g., Skype, Viber and WhatsApp), through online social networking (e.g., Facebook and Twitter), or through emailing. Rainie and Wellman (2012: 126) note, in this regard, that American teens refer to texting as "conversations" rather than as "writing." These virtual forms of quiet and silent communications can also complement noisy communications options in real space, which may take place either before or after noisy interactions. The Internet permits running away from noisy activities in public and semi-public action spaces, albeit differently than in real space circumstances: instead of people moving from the public arena into complete isolation, the Internet provides for interactive activities in the silence and privacy of virtual action space which take place within either silent or quiet real space settings of the Internet user.

The decline in the significance of the spatial dimensions of location and distance, as a result of the emergence of global social networks, may bear upon the sensing of place of users of online networking as well as on related processes of place production, because social life becomes removed from a local setting, in which social life is sustained within a routine spatial context. Thus, the geographically more dispersed social person-to-person ties may potentially be associated with increased placelessness (see, e.g., Dodge and Kitchin 2001; Wellman 2001). Halbwachs (1980: 134) proposed the terms *implacement* and *displacement* for social reactions to urban changes. By the same token, virtual person-to-person ties may imply a sensing of *spaceless places* (Ogden 1994: 715), referring to places that are not associated with any physical sensing nor with any known geographical addresses of the communicating persons.

Special types of social networks are those of professionals involved in creativity, looking for the " 'buzz', the interaction that generates creativity" (Jones *et al.* 2010: 99). In the R&D industry it has been assumed that such networks of creativity require face-to-face contacts, within the framework of high-tech industrial parks, thus bringing about the blooming of dedicated urban clusters for R&D firms located in relevant cities (see, e.g., Kellerman 2002; Wilson *et al.* 2013). There are those who believe that "buzz" may emerge also through online networks (Asheim *et al.* 2007), whereas others tend to be more pessimistic in this regard (Bathelt and Schuldt 2008). In yet another creative industry, Jones *et al.* 2010) were able to show the creation of some buzz in the American theater industry through a blog of New York theaters.

## Country data on online social networking

Table 8.1 presents a country breakdown of Facebook subscribers for the countries with a Facebook penetration rate among Internet users of over 90 percent in 2012. There are several striking elements in this table, as compared with the equivalent tables in the previous chapter (7.1–7.7), which presented the percentage adoption of a wide variety of commercial and semi-commercial Internet uses: first are the high levels of Facebook use percentages as compared to the percentage Internet users active in other, more commercial, action spaces; second is the identity of the leading countries in Facebook usage, which are neither the US nor Scandinavian countries; and third are the identities of the "lagging" countries, namely those with extremely low rates of Facebook subscription.

We will discuss these three trends, but before doing so it is important to note that the data in Table 8.1 seem reliable, as can be seen when comparing some of them to other equivalent data. Thus, the data source for Table 8.1, Internet World Stats (2013), reported for the US in 2012 a percentage usage of 68, whereas PewInternet (2013b) noted for the same year 67 percent. By the same token, Internet World Stats (2013) quoted for Indonesia in 2012 a rate of 93 percent for Facebook users only, whereas the Ipsos (2012a) survey asking interviewees about their "visit social networking sites/forums of blogs" found for Indonesia, for that same year, 91 percent. Internet World Stats (2013) further stated for Germany in 2012 a penetration rate of 38 percent, while the EU (Eurostat 2012b) mentioned for Germany for the same year a rate of 42 percent for the percentage Internet users "posting messages to social media." On the other hand, there are also some gaps in the data, which seem significant, such as for Montenegro, the leading country in the EU for the adoption rate of Facebook among Internet users, for which Internet World Stat reported 93 percent for 2012, whereas Eurostat (2012b) reported merely 77 percent for those Internet users "posting messages to social media." Such gaps may be attributed to multiple Facebook accounts managed by single individuals, a trend which deserves some closer attention.

Table 8.1 presents numerous countries for which the percentage of Facebook users is 100 percent and over. These percentages seemingly relate to multiple Facebook subscriptions maintained by the same persons. This seems to be typical to small countries, as well as to developing ones. This multiple subscription might probably relate to domestic cultures, in which there has rapidly developed the habit of establishing fake identities over online social networks, or possibly that domestic cultures do not permit open and free networking among people, notably from opposite sexes. We noted such trends already before

*Table 8.1* Facebook users as percentage Internet users in leading countries (over 90 percent) 2012

| Country | Percentage |
|---|---|
| Monaco | 118 |
| Iraq | 115 |
| Cambodia | 112 |
| Sierra Leone | 111 |
| Botswana | 109 |
| Central African Republic | 108 |
| Gibraltar | 105 |
| Mayotte | 104 |
| Jordan | 103 |
| Gabon | 102 |
| El Salvador | 100 |
| Guadeloupe | 100 |
| Belize | 100 |
| Grenada | 100 |
| Marshall Islands | 100 |
| Palau | 100 |
| Saint Barthélemy | 100 |
| Samoa | 100 |
| Nicaragua | 100 |
| French Guiana | 100 |
| Isle of Man | 100 |
| Saint Kitts and Nevis | 100 |
| Nauru | 100 |
| Turks and Caicos Islands | 99 |
| Democratic Republic of the Congo | 99 |
| Somalia | 98 |
| Chile | 97 |
| Costa Rica | 94 |
| Ethiopia | 94 |
| Papua New Guinea | 94 |
| Indonesia | 93 |
| Montenegro | 93 |
| Curaçao | 93 |
| Guatemala | 92 |
| Honduras | 92 |
| Mexico | 91 |

Data source: Internet World Stats (2013).

regarding Morocco (Hassa 2012) and Thailand (Hongladarom 2011), leading, though, to conflicting adoption trends. Thus, in Morocco in 2012 the percentage Internet users in the population reached 51 percent, whereas Facebook users among Internet users reached only 31 percent, attesting to a probable lower actual subscription rate for Facebook. In Thailand the opposite was true: whereas the percentage Internet users in 2012 stood at 30 percent, the share of Facebook users among them reached 88 percent, attesting to a possible wide use of fake identities.

The list of leading countries in the adoption of Facebook, as the most popular platform for online social networking, is completely different than the list of the repeatedly leading countries in the use of commercial and semi-commercial action spaces, which we noted in the previous chapter, namely the US and the Nordic countries. Even if we assume a wide use of multiple identities by Facebook users, it still seems that the online social action space is more popular in developing countries in all continents as compared to the virtual social action spaces in developed ones. Online social action space seems to be of special importance in small, notably island, countries, in which the small size of the population and the difficulties in the performance of physical mobilities to other countries are significant. Hence, the 2012 rates for Facebook users among Internet users in the US and in Nordic countries are in line with those presented for other more commercial online action spaces: 68 percent for the US; 59 percent in Sweden; 61 percent for Norway and for Finland just 49 percent (Internet World Stats 2013).

The leading countries in Table 8.1, Monaco, Iraq, Cambodia, Sierra Leone, Botswana, etc., were not on this list just two years earlier. Looking at the percentage of Facebook subscribers among Internet users in 2010, the five leading countries by the percentage Facebook subscribers among Internet users were: Indonesia (100 percent); Chile (90.6 percent); Venezuela (81.2 percent); Hong Kong (75.3 percent) and Turkey (69.0 percent) (Kellerman 2012a). The change in the list of leading countries in Facebook usage seems to attest to the fully global adoption of Facebook in recent years, as well as to the spread of the tendency for multiple identities in social networking in numerous countries, notably small and developing ones. This tendency for multiple identities may potentially decline in the future when the use of online social networking will stabilize, something that may potentially lead to social change in numerous countries with regard to values related to social networking at large. Thus, the list of leading countries in the use of social networking is probably about to continue to change in the upcoming years.

The popularity of online social networking in developing countries, as compared with their lagging in the adoption of other online action spaces, accentuates the difference between networking, on the one hand, and personal economic activity, on the other. Online social networking does not involve any financial costs for network subscription, and it is not related indirectly to the purchase of any products and services. The only cost involved is the general pay for the use of the Internet. For many users in developing countries who cannot afford the purchase and maintenance of Internet hardware and software, the access to the system is achieved through low-cost access offered by Internet cafés. The use of online social networking is, therefore, almost completely separated from the levels of national and personal economic development, whereas the use of online shopping, banking, government, etc., depends on personal incomes of the users, as well as on Web infrastructures of proper businesses. The only requirements for the use of online social networking are literacy and the little money for the use of publicly available Internet access. Furthermore, and from a social perspective, online social networking permits users to bypass social, cultural, religious and political taboos on social relations in real action space.

There are also numerous countries with extremely low rates of participation in Facebook. The lowest rate for 2012 was presented by China, with merely 0.1 percent, though we noted in the previous chapter a growing use of the more commercial online action spaces, such as for shopping, in China. This extremely low rate does not imply lack of interest in online social networking, but it rather presents the full banning of Facebook, until recently, by the Chinese government for political reasons, though domestic alternative online networking platforms are available. Another leading example of low participation is Russia and the previous Soviet Union nations with the following percentages: Russia (12); Moldova (17); Belarus (12); Ukraine (15); Kazakhstan (9); Uzbekistan (2); Kyrgyzstan (5); Tajikistan (4); Turkmenistan (4). It turns out that Facebook has a major Russian competitor in these countries, VKontakte, with some 215 million subscribers. Two other interesting examples are Japan (17) and Korea (25). In Japan, many networked people, notably youngsters, prefer the domestic system Mixi, in which subscribers make use of fake identities. In Korea too, preference is given to the domestic Cyworld network, as part of the general tendency there to prefer domestic Internet tools. Thus, in several parts of the world online social networking is as popular as elsewhere but domestic platforms are preferred over the international and American-made Facebook. Finally, the Facebook subscription rate for the Vatican City, among its Internet users, is also low and stood at merely 4 percent, given its special clergy population.

Interestingly enough, the share of Americans among Facebook subscribers worldwide in 2012 was less than 20 percent, declining rapidly from the 25 percent rate, just two years earlier, in 2010 (Kellerman 2012a). Thus, the globalization of Facebook has been extremely fast if we consider its original introduction in the US only in 2004.

## Conclusion

The Internet permits individuals to maintain wide geographical and social action spaces, and it further facilitates the management of wide social contacts, as well as the sharing of contents with them. These traits of online social networks stem from the very constitution of the Internet as an informational action space, which is not the case for real action space per se. Internet social communications is both locationally and temporally flexible, since it does not require the synchronous attendance of the communicating parties, as is the case for most of real space social relations.

Social networking, within the wider networked information economy, may be viewed as individual actions, coordinated with those of other people's individual actions, and aided by the distributed mechanisms provided by information technology through proper platforms. Jointly, these elements may bring about social ties of numerous types and forms. Networked individuals present a new sense of personal, albeit connected, autonomy, expressed also through increased personal mobilities (see also Kellerman 2012a).

The very nature of the social relations developed online through social networks, as well as their potential impact on social relations in real space, has been widely debated in recent years. This debate has taken place in parallel with the massive adoption of social networking by Internet users globally, so that societies worldwide are still in search of balances between real and virtual social action spaces and relations. It is assumed, though, that repeated virtual communications among parties may frequently lead to face-to-face meetings of the parties.

We may quite safely assume that by 2013 some three-quarters of the Internet users worldwide were engaged in social networking over the net. Thus, the barrier-free, constraint-free (in most countries) and cost-free action space for virtual social networking has exhibited a tremendously fast adoption. However, the very use of social networking over the Internet does not automatically imply globally stretched networks attended by a large number of subscribers, and this since local communications are still a preferred networking mode, notably by teens.

In the process of development of the Internet into a second action space for individuals, social networking has turned out to be by far, and at the global scale, the most popular activity to take place over the virtual action space offered by the Internet. In developed countries, online social networking is as popular as other activities taking place online, whereas among Internet users in developing countries (a minority of the total adult population in most cases), it is even more popular than in developed ones, and this trend is accentuated even further in developing countries by the weak uses of the Internet for other activities, mostly those of a more economic nature.

# 9 Darker actions over the Internet

This chapter will focus on some of the darker sides of the Internet and its action spaces, legal darker sides, as well as illegal ones. Some of the operations that take place over the Internet, which are considered as dark, are actions such as identity theft, which are performed by a small number of users who attempt to benefit themselves while damaging numerous others, without their engagement in any specific virtual activity. Thus, the effect of such dark uses might be extensively wide, reaching and affecting numerous innocent users in many systems and countries. Other darker sides of the Internet consist of measures instituted by organizations rather than by individual users, in an attempt to avoid actions by others, such as censorship imposed by governments in order to avoid or in order to direct Internet actions by their citizens. Darker activities may, thus, be performed by governments, groups and individuals. The darker side of the Internet constitutes the other side of the "open code," which was highlighted, at the time, as the most important feature of the Internet (Lessig 2001). Hence, the very existence of free flow of information over the Internet may be used negatively via cybercrime, or it may even be blocked through cyberobstruction, and we will generally outline these two terms in the next section. This general discussion of the darker sides of the Internet will be followed by more particular explorations focusing on: surveillance, identity theft, censorship, hacking, pornography and online gambling. These general and specific discussions will then be complemented by the presentation of some country data on cybercrime and cyberobstruction.

Among the numerous types of cybercrime and cyberobstruction, the choice of activities for discussion in this chapter reflects two dimensions. The first dimension includes crimes or actions that attempt to block or to harm the use of the Internet for any of the action spaces described in the previous chapters. Such crimes or actions can be divided into three types by their performers. First, there are those crimes that are performed mainly by individuals, represented in this chapter by surveillance and identity theft. Second are negative activities that are performed by state governments, presented here through censorship, and third, there are crimes and negative activities performed varyingly by state governments, groups and individuals, and represented here by hacking.

The second dimension of the negative activities chosen for this chapter relates to activities which constitute action spaces by themselves, and to which some negative and sometimes criminal aspects have been attributed: online gambling and pornography. The first of these two activities, online gambling, imitates real space casinos, and is, therefore, a virtual action space in similar ways to the numerous ones we discussed in Chapter 7. The second activity, pornography, involves mainly the presentation of visual and textual information rather than being a virtual action space by itself, but it is based on an extensive production industry for the websites in real space, side-by-side with other related activities in real space, such as prostitution. In addition, watching pornography may bring about immediate actions by viewers, such as masturbation.

There are numerous other dark activities that take place over the Internet, activities which are mostly of an informational nature, and will not be discussed, therefore, in this chapter. The major activity in this regard is hate and racism. For 2011 it was found that over 14,000 hate sites, blogs and social networks were in operation over the Web, involving the so-called "information launder-ing" (Klein 2012: 427).

## Cybercrime and cyberobstruction

Cybercrime was broadly defined as "a wide range of activities relating to the use of information technology for criminal purposes" (Kraemer-Mbula *et al.* 2013: 543). As such, numerous criminal activities which have been performed in real space much before the introduction of the Internet have found their way also into cyberspace, such as surveillance, racism and gambling. Other cybercrime activities might be viewed as novel ones and stemming from the very nature of cyber-space as a sophisticated web of information technology platforms and software. Striking in this latter regard is hacking or the creation and distribution of malware. The motivations for cybercrimes are no different than those for crime in real space: money; power; satisfaction and amusement. Some of the technolo-gical tools used for cybercrime may turn out to be useful for positive and legal purposes, such as GPS technology which has turned out as being most helpful for destination finding, as well as for LBS services, but this same technology might further be used for personal spying activities (Dobson and Fisher 2007).

By using the term cyberobstruction we refer to attempts, notably by govern-ments, to obstruct users from accessing certain materials over the Internet. Such obstruction takes similar forms to the obstruction of personal mobility in real space. For instance, Internet censorship that blocks the very communications ability of Internet users and/or users' access to certain websites and communica-tions platforms is similar to roadblocks in real space. Another example is laws and regulations forbidding the use of certain materials that are accessible over the Internet, but not blocking the ability of users to access them. Such attempts are similar to no-entry signs on physical roads. This kind of cyberobstruction is taking place when laws are instituted in order to prohibit access to hard pornography materials, for instance the British "dangerous pictures act" (Wilkinson 2011).

Whereas the definition of cybercrimes that are similar by their very nature to those in real space is quite clear, the definition of the so-called criminal activity that is Internet-specific is more problematic. Thus, hacking is considered a crime in many countries, but side-by-side with its prohibition, governments may use hacking as a military weapon, such as the speculation that the US and Israel developed and spread hacking software against the Iranian nuclear efforts. Yet from another perspective, the hacking of computer systems for military purposes may divert defense activities to non-killing ones.

The very use of cyberobstruction measures by governments is a matter of ideology. Some would claim for the very authority of governments to defend their regimes at the expense of free access to and spread of information by their citizens, whereas others would praise those citizens who would try to circumvent governmental censorship in an attempt to foster democratic values. By the same token, some would agree with the prohibition of viewing hard pornography materials whereas others would claim that such regulations might hurt values of free speech (Wilkinson 2011). Still, the use of some kind of cyberobstruction of children's exposure to pornography over the Internet might be considered by others as a needed educational tool, and, by the same token, the use of children for pornography is widely considered a crime.

## Surveillance

Rainie and Wellman (2012) differentiated among several types of monitoring or "-veillance" activities over the Internet: surveillance; coveillance and sousveillance, and we will outline these three types in the following paragraphs. When combined they jointly imply a potential and frequently effective complete loss of privacy of individuals, in all their virtual action space activities: networking personally, professionally, or when acting as customers of any online service. "Veillance" of all types will emerge as a leading concern in the future development of the Internet as a second action space, and it may affect also actions in physical space being monitored through GPS, cameras, and other devices. The open code nature of the Internet and its use by non-innocent groups, whichever defined, may continuously call for the breaking of privacy. The breaking of privacy may also be initiated by governments, as well as by commercial companies interested in the habits and preferences of Internet users. Furthermore, individuals may be involved in the breaking of privacy of fellow Internet users, when their curiosity is stimulated by the availability of personal information of fellow Internet users.

The common thread among the numerous types of "veillance" which we will note is that despite the seemingly autonomous nature of personal online action spaces, devoted to social, economic and any other uses, individuals' communications may be recorded and/or used by others. Commercial companies may record Web surfing of Internet users in order to channel advertisements directly to target potential clientele, and similarly, the location of mobile telephone users may be identified through GPS and other technologies for LBS advertisements. In addition, personal information may be disclosed to other users, and governments may monitor communications and Internet activities. These rather diversified monitoring activities present numerous motivations by those performing them, whether commercial (by companies), political (by governments) and social (by individuals). Such monitoring may imply risks for individual users, notably for those engaged in intimate or political social networking. Thus, the exposure of communications to unknown others, may hurt notably those people who are involved in social networking activities, and wishing to express through its use their autonomy of operations.

"Surveillance" refers to monitoring activities operated by governments and organizations, notably of online social networking, as well as of Internet surfing and of the Internet action spaces of individuals in general. Surveillance has been widely discussed elsewhere (see, e.g., Murakami Wood and Graham 2006), and will, thus, be outlined here only briefly. The largest community of Internet users, the Chinese one, operates systematically and openly under governmental surveillance. Following the September 11, 2001 terrorist attack in the US, the US government introduced selected surveillance as well, monitoring its citizens who comprise the second largest community of Internet users. This domestic surveillance was followed later by massive international monitoring, the PRISM project, which was recently revealed. Such monitoring activities are carried out by other governments, as well, presenting all regime systems, whether for security or for political motivations (see Morozov 2011). Commercial companies are engaged in commercial monitoring, and these focus on personal traits, habits and shopping preferences of individuals in order to target potential customers (see Zook and Graham 2007b). Another kind of surveillance was termed "geo-surveillance," referring to the illegal monitoring of the spatial locations and movements of people, through illegal hacking via GPS, as well as through other geospatial technologies, normally used by the police for the monitoring of criminals (Crampton 2007).

"Coveillance" relates to the search by individuals for personal information of fellow Internet users, notably within social networks, and when this type of "veillance" is performed by teens it is sometimes called "creeping" or "stalking." Facebook announced recently the introduction of a dedicated search engine in this regard. On the other hand, network subscribers try increasingly to protect or securitize their personal information by limiting access to their personal profiles and posts (see Pinkerton *et al.* 2011), thus, in fact, restricting the social action space of networked Internet users.

"Sousveillance" refers to the contemporary ability of individual users to access confidential correspondence and communications by governments and organizations, made possible through massive leaking efforts, for instance those recently made public by organizations such as Wikileaks.org. Such information may potentially provide individuals with enhanced political power. "Souveillance" may be viewed as constituting a kind of contra attempt by individuals and organizations to the massive governmental surveillance of individuals, thus presenting the possibility that all Internet communications by all users, including governments, will become transparent to unwanted parties.

## Identity theft

Whereas "veillance" and pornography are partially legal and widely debated for the dos and don'ts associated with them, identity theft seems to be considered globally as cybercrime, though in different ways in several countries. Thus, in the US and Canada identity theft has been considered a serious crime, whereas in Europe it has been classified as fraud (Acoca 2008). Identity theft has become the most popular and most growing criminal action space, among all virtual as well as real action spaces.

The motivation for identity theft is normally financial, with criminals attempting to acquire funds illegally, in what is considered as "the cutting edge of global criminal activity" (Kraemer-Mbula *et al.* 2013: 541). However, identity theft can be made for a variety of other reasons as well, such as medical and legal needs. Identity theft can be performed through malicious software ("malware"), or through the "phishing" of emails and websites in numerous ways (Acoca 2008). Governmental reporting for the US revealed that in 2010 some 7 percent of American households experienced the victimization of at least one of their members by identity theft (NCJRS 2013). Organizational and individual Internet users have, therefore, to invest attention and funds for the protection of their own access to the system, and their activities in it, whether using fixed or mobile devices. These needs for security measures for the prevention of identity theft have brought about the development of a wide industry of security software.

## Censorship

On July 5, 2012 the United Nations Human Rights Council declared that it "*affirms* that the same rights that people have offline must also be protected online, in particular freedom of speech, which is applicable regardless of frontiers and through any media of one's choice" (UN 2012). However, despite of this declaration agreed by 47 member countries, in his recent regional review of the global geography of Internet censorship, Warf (2013: 71) noted that, "most governments seek to appropriate the economic benefits of information technology without paying the political costs of enhanced democracy." Thus, the Internet is considered as a needed medium by almost all countries, but simultaneously this is also coupled in many of these countries with some kind of censorship. The motivations for Internet censorship are multiple, ranging from political repression of all or of a minority of dissident citizens, through religious controls of the viewing of materials, notably sexual ones, to cultural oppression of ethnic and other minorities. Whereas from a democratic perspective all these motivations may be considered as negative, there is another type of Internet censorship, which might be viewed by some as a positive one, namely the attempts to protect intellectual property from illegal downloading, notably of music and movies.

The very operation of censorship involves rather numerous and diversified measures, such as: limits on the number and width of international Internet channels, so that their contents can be filtered and controlled; automatic delays in email traffic, thus permitting some message screening; manual surveillance and tracking of individuals; and severe punishment for attempts to violate censorships. On the other hand, the open and sophisticated structure of the Internet makes it difficult to impose full and absolute censorships, so that interested individuals and groups frequently attempt to bypass the imposition of censorship, using a variety of technological and organizational means, such as periodical changes in website names; lack of congruence between website names and titles, on the one hand, and their contents, on the other; as well as more sophisticated means, such as the use of foreign proxy servers, and the development of circumventing software tools. Thus, censoring states and some of their citizens find themselves in a continuous fight on degrees of Internet freedom. The imposition of governmental censorship implies restricted action spaces for individuals, notably as far as social ties are concerned, sometimes limiting mostly international ones, as is the case in China, and sometimes even restricting domestic ones, as is the case for Iran.

## Hacking

Internet hacking constitutes computer network attacks (CAN) (Liff 2012), which may be executed by individuals, groups or governments, aiming at a wide disruption of emailing and websites activity. Computer hacking began much before the introduction of the Internet, back in the late 1950s (Betz 2012), but it has flourished with the vast possibilities over the Internet. Hacking includes actions such as the installation in the Internet of types of malware (malicious hardware) and botnets/distributed denial of service (DDoS) (combinations of malware, notably for the disruption of networks). These actions have, thus, brought about a flourishing computer security industry, attempting to provide protection against hacking activities.

Hacking has been performed by individuals, including frequently students of computer studies, for fun, for a demonstration of sophisticated knowledge of information technologies, or in attempts to disrupt the Internet activities of some specific companies or governments. This latter aim of the disruption of company and governmental Internet activities has been also the aim of organizations, notably that of "Anonymous," which has attacked at times, the music industry, credit card companies, religious organizations and governments, justified by changing motives. Governmental uses of hacking, within the so-called *cyberwar* attacks, include the Russian attack against Estonia in 2007 and the two-way attacks by Georgia and Russia in 2008. The Stuxnet attack on Iran in 2010, attributed to the US and Israel was not considered a CAN by Liff (2012), since its aim was not to disrupt Internet systems, but rather to disrupt a real space target, the centrifuges of the Iranian nuclear program. Cyber commands, aiming at cyber attacks and cyber defense, have, thus, been established in numerous armies.

## Pornography

The Web has facilitated access to pornographic pictures, films and video clips at an unprecedented extent, making such access as easy as accessing any other visual materials available on the Web. For 2010 it was estimated that some 4 percent of the million leading websites were sex related, and that some 13 percent of the Web searches by then were for erotic content (Forbes 2011). Whereas "velliance" activities are viewed as infringing individual privacy, pornography has been considered in diversified manners by countries (see Zook 2003). For some countries the production and posting of Web pornography is permitted as part of a wider provision of freedom of expression, while for others such activities have been considered as constituting desecration of public morality, either by the very production of pornographic materials or by their posting on the Web. In some other cases, even the viewing of pornographic materials is considered illegal. The issue of pornography is, therefore, a matter of geography even for virtual space, and its constituting cybercrime is, hence, country specific, notably because a complete prohibition of the viewing of pornographic materials may be assessed as an obstruction of free speech and thought (see, e.g., Easton 1994). Thus, in 2010 some 68 countries had specific laws prohibiting child pornography in some way, whereas 92 other countries did not have such legislation (ChartsBin 2010). Furthermore, some specific types of extreme pornography may be prohibited, even for viewing, in several countries, notably children pornography, as is the case in the UK (Wilkinson 2011). The pornographic virtual action space is, therefore, highly contested, in all its aspects: production, posting and viewing.

## Online gambling

Just a year after the introduction of the Internet in 1994, the first online casino gambling games emerged in 1995, albeit by then without real money involved (Wilson 2003; Wood and Williams 2007). Fifteen years later, in 2010, some 2,679 Internet gambling websites were identified, owned by some 665 companies, and offering 865 online casinos (Stewart 2011). This global online gambling industry was estimated to generate revenues standing at $12 billion per annum (Valentine and Hughes 2011).

Online gambling has turned out to be an attractive virtual action space, side-by-side with casinos located in real space, and for numerous reasons, which are related to both the demand and the supply sides of the industry. From the perspective of gamblers, a leading advantage of online gambling is its availability from anywhere, and at anytime, thus bypassing the legal total prohibition to construct casinos in numerous countries, or its restriction to specific regions or cities, as is the case in many other countries. In addition, online gambling provides ease of access, notably since it is accessible from homes. Online gambling, thus provides for much privacy in its use, and hence the high percentage of women who make use of online gambling systems, sometimes pretending to be males. Furthermore, online gambling does not require the use of cash money or material chips, it often provides multilingual service, and it is believed to offer better payout rates, given the lower construction and operation investments required by online gambling companies, as compared with the equivalent ones for real space casinos. Furthermore, competition among gambling websites is high, as it is easy for players to move to competing websites (Wilson 2003; Wood and Williams 2007; Valentine and Hughes 2011).

For gambling companies, or the supply side of online gambling, as we mentioned already, it is cheaper to establish and maintain online casinos than those in real space. However, real space casinos may build over the years a reputation and trust by their customers, something which is more difficult and time consuming for online gambling websites to establish and promote. On the other hand, though, gambling websites are available at any time to a much wider potential clientele than real space casinos. Gambling websites sometimes seem to be located in the US, while actually operating from offshore countries (Wilson 2003).

The legal concerns regarding online gambling are, foremost, similar to those concerning real space casinos, notably people potentially becoming addicted to gambling, coupled with family and personal problems. Further legal concerns relate to the legality of the monies invested by players in gambling and the payment of proper taxes by both gambling websites and their winning customers (Wilson 2003; Wood and Williams 2007; Valentine and Hughes 2011).

Online gambling is similar to online pornography in that its status of legality differs from country to country. As far as online gambling is concerned, legality may mean the permission to locate websites within countries, and/or the permission for citizens to use online gambling wherever located (Wilson 2003). In 2010, some 85 countries legalized Internet gambling (Stewart 2011). Since its inception, the preferred location of websites for online gambling have been small and offshore countries in the Caribbean and South America, such as Antigua and Costa Rica, side-by-side with similar countries in Europe, such as Gibraltar. In more recent years, though, larger European countries, such as the UK, Italy and France, as well as the US, Canada and Australia, permitted partially the location of gambling websites within their territories (Wilson 2003; Stewart 2011).

## Country data on cybercrime and cyberobstruction

Table 9.1 presents the combined percentage of leading countries in their global share in five major types of cybercrime: malicious code; spam zombies; phishing website hosts; Internet bots and hacking attack origin. The data go back to 2009, the latest year for which such data were released by Symantec, the producer of the widely used antivirus software Norton (Symantec 2010).

The country which led in malicious Internet activity in 2009, both absolutely and relative to its share among Internet users was the US, a country which has led also in online shopping (Table 7.7), and the country which can probably be considered as the pioneering one in the emergence of online action space in

*Table 9.1* Country share in global malicious Internet activity and among Internet users
2009

| Country | Percentage malicious activity | Percentage Internet users |
|---------|-------------------------------|---------------------------|
| US | 19 | 14.2 |
| China | 8 | 22.5 |
| Brazil | 6 | 4.4 |
| Germany | 5 | 3.8 |
| India | 4 | 3.5 |
| UK | 3 | 3.0 |
| Russia | 3 | 2.4 |
| Poland | 3 | 1.3 |
| Italy | 3 | 1.7 |
| Spain | 3 | 1.6 |

Data sources: Malicious activity: Symantec (2010), see also Kraemer-Mbula *et al.* (2013). Internet users: Calculated from data in CIA (2013).

general. The US is further the country with the declared highest level of Internet freedom (Warf 2013). It seems, therefore, that there is a price tag to Internet freedom, as a societal value, as well to leadership in technological innovation and its wide adoption, notably as far as the Internet is concerned. These combined characteristics of the American society have been negatively used by cybercriminals. There are several European countries, both Eastern (Russia and Poland) and Western (Italy and Spain) in which the share of malicious Internet activity was significantly higher than their national share in Internet usage, potentially attesting to some permissive atmosphere in this regard.

Censorship distribution has also been calculated for countries worldwide. The 2011 scores for the Press Freedom Index were calculated by Reporters without Borders (RWB) (2012), based on their and on their partner organizations' assessments of some 44 criteria, covering all the media, including the Internet and the degree of freedom there for its users. Under the assumption that censorship is imposed similarly in both real and virtual spaces, the Press Freedom Index might serve as a surrogate for pure Internet freedom.

As Table 9.2 shows, about one-third of the world Internet users enjoyed only minimal censorship in 2011. This includes North America, the Nordic countries and Germany, whereas about one-quarter of the Internet users, mainly those living in China, Iran and Saudi Arabia, suffered from the severest levels of censorship during that year (see Reporters without Borders 2013 and Warf 2013: Figure 3.1). Some additional 44 percent of the global Internet users used the system under mid-levels of censorship. The countries which we noted in Chapter 7 as those leading in the mass adoption of commercial action spaces have also been those which enjoyed the highest levels of Internet freedom. However, this correlation may change in the upcoming years with the growing adoption of online commercial action spaces by Chinese users, who may simultaneously continue to use the Internet under the severest levels of censorship. It remains to be seen whether it will be possible for new and pioneering Internet action spaces

*Table 9.2* Global percentages of Internet users under differing levels of press freedom 2011

| Score | Percentage of Internet users |
| --- | --- |
| 0–9 | 31.9 |
| 10–19 | 16.2 |
| 20–49 | 20.3 |
| 50–79 | 7.1 |
| 80–115 | 24.4 |

Data sources: Score: Reporters without Borders (RWB) (2012). Percentage of Internet users: Warf (2013).

to be invented and initially adopted in countries that impose severe Internet censorship, or alternatively, that such countries can only follow free countries in which new online action spaces are invented and initially adopted.

## Conclusion

Darker Internet activities may be performed by governments, groups and individuals. The joint effect of darker activities by all potential actors of such activities imply the limiting of free, full and safe access and action on the part of innocent Internet users worldwide. This is true for the case of identity theft, the leading cybercrime, which attempts to benefit some actors at the expense of the damaging of others. It is also true for other dark sides of the Internet, consisting of cyberobstruction measures which are instituted in an attempt to avoid actions by others, such as censorship imposed by governments in order to avoid, or in order to direct, the Internet actions by their citizens. Similarly, this is further true for hacking, whether performed by individuals, groups or governments.

The problem of potential limits to personal action, coupled with questions of potential limits of governmental actions, have accompanied the long development of societies in real space, and this dilemma has been imported into virtual space as well, with obvious special forms characterizing informational and global virtual space. The debate and the fight between the acceptable and the unacceptable, between the permitted and the forbidden, notably with regard to pornography and online gambling, will accompany the future of Internet action spaces, presenting differences among countries worldwide on both political and cultural bases.

# 10  Conclusion

This concluding chapter will first present summaries for all the previous chapters. We will then outline several concluding perspectives, referring, first, to both action and actors in the light of transitions in spaces of action, and, second, referring to the changing significance of time at times of transforming action spaces. The chapter will conclude with some thoughts on potential future trends. Can people live easily in the two spaces, the real and the virtual ones? What is the potential future of these two spaces? What is the technological horizon for future developments of virtual space?

## Book summary

The first part of the book was devoted to the Internet as a platform for virtual action space. Thus, back in Chapter 1 we reviewed and highlighted cyberspace and the Internet. First, we defined cyberspace, focusing on its nature, and, second, elaborated on its two classes of information and communications spaces, and on its cognition as a virtual entity. We then moved to a third elaboration, comparing real and virtual spaces in their basic dimensions of organization, movement, and users, followed by a comparison of the practical relations of these two spaces. These three rather basic and rather general discussions of cyberspace paved the road for a brief presentation of the Internet per se, its foundations and its characteristics as an open and universally accessible information system. The Internet has become increasingly accessible in recent years through mobile information technologies, which we described and assessed in terms of the more extensive actions that they permit in their constitution as mobile systems. We ended our introductory discussions on cyberspace and the Internet by adding to the visual perspective normally attributed to the Internet some comments on the auditory geography of the Internet, as compared to that of real space. Following these discussions of the foundations of the Internet and its access we paid some attention to the notion of action space and presented the classical definitions for action and activity spaces, assessing them vis-à-vis the Internet.

A repeated theme in the first, introductory, chapter was the relations between real and virtual spaces, and we noted several such relations. We discussed the practical integration and convergence of cyberspace with real space, namely: the two spaces as being interfolded, side-by-side with cyberspace as being embedded in real space; the organization of cyberspace in similar forms to those known for real space; and, finally, cyberspace as representing real space. This variety of relations between the two spaces points to rather complex relations between them, a topic we further highlighted in Chapter 2, devoted to theoretical perspectives of the Internet as a second action space. This diversity of relations between the two spaces may also point to a still emerging and as of yet unsettled development of the Internet as an action space, and thus its interrelationships with the more traditional real space being still in the making.

Three possible relations between real and virtual spaces were recently proposed by *The Economist* (2012): the digital world reshaping the physical one; the real and the virtual worlds as separate entities from each other; and the physical world as shaping the virtual one. The discussions in Chapter 1 attempted to show that so far all these three possibilities are true, and their relevance to different spheres of the virtual world was outlined in later chapters.

In Chapter 2 we delved a bit deeper into the nature of real and virtual spaces, attempting to explore some basic meanings, significances and workings of these two spaces, notably from the perspectives of their constitution as both social and action spaces. Four perspectives were offered in this chapter: attributes of real and virtual social spaces; experiencing real and virtual social spaces, notably under the assumption that virtual space has turned into action space; social forces in the emrgence of virtual action space and time geography aspects of virtual action space.

It was shown in Chapter 2 that the rather numerous attributes, interpretations and metaphors which have been suggested in the literature over the recent decades for real social space are all relevant for virtual space as well, though sometimes in different ways. This relevance of the attributes for real social space for virtual social space has prevailed despite the lack of geographical scales, such as city and region in virtual space. The issue of the experiencing of virtual action space was discussed in Chapter 2 vis-à-vis Lefebvre (1991; developed further by Harvey 1989 and Soja 1996), whose notions of space types have seemed to be challenged by the emergence of the Internet as an action space. It was argued that virtual action space is a novel entity and somehow differing from real space in its experiencing, being more flexible in both its very construction and in moving through it.

The social meanings that may be attributed to the construction process of virtual action space were highlighted through structuration theory (Giddens 1990), which accentuates the vicious cycle between human agency, on the one hand, and social structures, on the other. It was argued that information technology has served as a mediator beween human needs and actions as reflected and performed in real space, on the one hand, and business profit motives, on the other. These two forces, jointly and gradually, have shaped virtual action spaces through information technology. Distanciation has reached its utmost in globally accessed virtual action space, whereas time-space compression has been amplified to unprecedented levels in the global availability and use of cyber-space. Finally, virtual action space was assessed by the time-space prism and time-space constraints proposed at the time by Hägerstrand (1970, 1973, 1975). It was argued that the time-space prism does not exist anymore at times of double space action by individuals, notably with the growing use of mobile broadband for accessing the Internet. Furthermore, several of the time-space action constraints have either changed or have diminished for virtual action space.

The theoretical discussions in Chapter 2 tended to present the seemingly maturing virtual action space as an entity by itself, even if its hardware and its very access by users are rooted in real space. However, as Jordan (2009: 182) noted: "what we once called 'virtual' has become all too real, and what was solidly a part of the real world has been overlaid with characteristics we thought of as belonging to the virtual." With the sophistication of both the very access to virtual action space and the actions performed within it, the very status of real and virtual action spaces is still being challenged, as well as the relations beween them, as we noted already in Chapter 1. Thus, some new action processes and patterns may emerge in the near and/or far future, possibly bringing about new rounds in the reshaping of our activity customs and patterns within the two spaces, as well as in the numerous dimensions of these two spaces.

Following the discussions in Chapters 1–2 we could portray, in Chapter 3, the structure of the Internet action space for average individual users of the Internet. Internet users comprised some three-quarters of the population in developed countries and about one-third of the population in developing ones in 2013. All in all, there are two leading trends which characterize the Internet at large and its constitution as an action space in particular: the dominance of the US, which is coupled with the dominance of major commercial players in its operation. This latter dominance implies that the technical, visual and operational structures and envelopes of the Internet are quite similar worldwide, while some particular pat-terns of their uses and applications may depend on domestic cultures. Such domestic cultures may diminish in their significance in years to come, if we assume the continuing growth in digital globalization, so that uses of the Internet may become even more similar worldwide.

With 42 percent of all the websites worldwide hosted in the US and almost 60 percent of them registered there, there is a good chance that about one half of individual accesses and Web browsing sessions by individuals worldwide do involve some American component in them. Moreover, even if a domestically registered and hosted website is accessed, there is still some chance that this seemingly inland connection between users and websites will be made actually through the US, which is the most widely connected country, through both trans-Pacific and transatlantic cables. It is almost universal that the enveloping and major operational tools employed by individual Internet users are once again American and, thus, the same all over the world, including foremost Windows; Word and Google. Similarly there emerge American dominance patterns in smartphone operational systems, notably Android and iOS. It remains only for the browsers market to present a slightly wider market beyond the two major players, Explorer (Microsoft) and Chrome (Google). There is further a large share of global Internet users that subscribes to geographically wide-ranging social networking system, either Facebook, which may soon include one half of global Internet users, or other popular networks, such as Twitter.

This almost full standardization of the technical, visual and operational structures and envelopes for Internet operations by its users, reminds one of the urban landscapes of real space cities. These too have similar elements worldwide, consisting of major streets and highways, coupled with more minor paths, as well as high and low buildings; stores at the street level; street lights, etc. These similarities in the general components and structures of urban landscapes permit the use the same digital navigation systems (GPS) everywhere. By the same token, website developers need to care only for language choice (or to direct users to proper translation tools), whereas the technological infrastructures are mostly standardized, so that the Internet action space and its ways of operation and use continuously move towards full standardization. Furthermore, Internet users can make use of the system using computers in any country without knowledge of the domestic languages, since the looking and structures of Windows and Google screens are similar worldwide.

The societal breakdown of Internet users and uses seems to be quite complex. Whereas the exposure of women to the Internet is lower than that of men in most countries, either marginally or significantly, their preferred personal uses of the system differ from those of men, with an accent by women on networking. As compared to other technologies, the Internet is a technology which has been widely adopted by children and adolescents, and even more so by young adults, again with a high popularity of networking among them. However, the penetration of the Internet by geographic regions points to higher levels of adoption in urban areas, mostly because of the differences in human capital between urban and rural areas.

In Chapter 4 we moved to the second part of the book devoted to human needs and the Internet. The basic human needs of physiology; safety; love/belonging; esteem and self-actualization have recently come to share their fulfillment between physical and virtual spaces, brought about by the contemporary Internet and communications technologies, which permit activity in a double space. The role of virtual space in this regard grows along the hierarchy of these needs, moving from secondary importance for the most material needs, the physiological ones, to a major importance for the more abstract ones of knowledge acquisition for self-actualization at the top of hierarchy of human basic needs.

This growing role of virtual space along the hierarchy of basic needs is expressed in hierarchical relationships between the two spaces along the hierarchy of human needs. Thus, in the lower levels of the hierarchy, virtual space offers mainly complementarity to physical space, which constitutes the crucial arena for the satisfaction of the most basic needs, moving to competition with physical space in the higher levels of the hierarchy, and eventually substituting physical space at the highest level of the hierarchy. These rather diversified and simultaneous relationships between the two spaces regarding the fulfillment of human needs are significantly different than the singular relationship of complementarity proposed at the time regarding the relationships between physical and virtual mobilities.

A special relationship of escape from physical to virtual space has been brought about by social networking, and this kind of escape is similar to the escape from daily routines offered by tourism in real space, through temporary travel to other places. This virtual escapism through networking is new, though already widely and swiftly adopted. The swift global growth in online networking was discussed in greater detail in Chapter 8, but it is still too early to speculate on future routines that may emerge among individuals with regard to such escapes. Further with regard to the future, it does not seem real to foresee that virtual space will offer any spatial exclusivity for the fulfillment of yet new and unforeseen human needs.

Another human basic need, curiosity, was the focus of Chapter 5. Three spatial dimensions of curiosity were proposed, namely trigger, objective and means, and all the three have been presented as undergoing some major changes in the information age, changes that are dominated by the emergence and wide availability of the Internet, at least in developed countries. Thus, virtual websites and their attractive design, side-by-side with growing networking, have provided people with extensive triggers for curiosity. Curiosity as an objective has also undergone a major change through the growing possibilities for individuals to move in physical space as tourists and scholars, and, thus, tour other places and countries. These new touring options have made it possible for people to be more curious about other places, and have made it possible for them to satiate such curiosities by, first, learning about them through the Web, followed by their physical visits in them.

It seems, though, that the major change in the information age regarding curiosity relates to the availability of new means for its satiation, mainly the Web and its enormous knowledge and information resources. These resources have become available to scholars, the professional "questors," as well as to laymen. The "electronification" of the availability of both tacit and codified knowledge, through emailing and the Web respectively, has brought about more complex research patterns through growing collaborations among scientists, rather than bringing about an increase in their productivity, as measured by the number of the publications by the scientific community at large. The dependence of scholars on virtual mobility, as compared with scholars' past dependence on physical locations within campuses, may attest to the recent statement claim that "it seems that [for them] the tyranny of space is being replaced by the *tyranny of mobility*" (Ferreira *et al.* 2012: 688). These trends in the work of scholars, dominated by the use of the Internet, are similar to trends in economic production at large which is becoming more and more dependent on email connectivity and on the use of websites for marketing.

Generally then, it seems that the nature of scholarly work, or the satisfaction of scholarly epistemic curiosity, has undergone dramatic transitions in the information age so far, in its turning of the Internet into a second action space, sometimes even into becoming the primary one. This has emerged for all of the major phases of research work: the search for relevant literature, mainly as digitized articles available through the Web, and partially also as digitized books; the very performance of research, notably cooperative, with emailing serving as the major communications platform; and, finally, the publication of research results, once again in digitized journals. Furthermore, the digitization of research resources has facilitated a locational flexibility for scientists, as compared with their previous fixed locations, in the close vicinity of academic libraries. Laboratories are still located in real space, but experiments and their results are usually available in digital formats, and thus are also being transmitted over the Internet.

An open question is whether non-professional questors in the Internet age, involved in their daily lives in prudential curiosities, experience too growing levels of curiosity as a result of the availability of the Web for the satiation of their prudential curiosity. This might well be the case, since ordinary people can afford it now to be curious more frequently and on the widest variety of topics, as they can satiate these prudential curiosities through the Internet, and most of the time quite swiftly so.

Another human need, the expression of which has received profound enhancement over the Internet is personal identity, which was the topic of Chapter 6. We indicated two actions regarding personal identities that are performed over the Internet: the presentation of identities, or the supply side of identities, and the search for identities, or its demand side. People may present some details of their internal identity through personal homepages or over social networks, and they might present their external-professional identity through the posting of their CVs, and/or through the construction of personal homepages. Internet users are, furthermore, frequently involved in fake and pretend presentations of themselves, notably in social networks. The other side of the coin of personal identities over the Internet relates to the search for personal information, either by presenters of identity or by non-presenters, searches made for professional, social or personal motives. Such searches are performed through dedicated websites for people searches, which normally provide paid-for services, but these searches can, obviously, be made also through search engines.

Despite the obvious differences between real and virtual spaces, discussed in earlier chapters, some basic characteristics and complexities of personal identity prevail in both spaces: in both of them people may present several and rather changing identities; in both of them people may present "true" and fake identities; in both of them there are identities presented by individuals, alongside their identities as presented by others; in both spaces people present their given/overt identities, complemented by their given-off/covert personal identities as perceived by fellow people; and, finally, in both spaces persons may possess a more stable identity as presented by their CVs in real space and by the equivalent personal homepages in virtual one. However, individuals may simultaneously present also changing identities through face-to-face and online chats.

The major element of personal identity which exists only offline in real space and which cannot exist by definition in virtual space is one's internal personal identity, though this internal conception and consciousness may be influenced, and may even be changed, by one's personal identity as presented over the Internet by oneself, as well as by others. Such change may occur since one's internal identity is generally sensitive to contextual-social circumstances.

The major dimensions of online personal identities reflect the unique nature of the relatively novel informational virtual action space: first, the distribution of personal identities of individuals has become potentially global, assuming that everybody's personal homepage maybe accessed by everybody else and from anywhere; second, and being the other side of this coin, people's personal identities can be accessed by everybody else without prior permission granted by the "owners" of the personal homepages; third, the availability to others of "third party" materials relating to one's personal identity, through dedicated websites for search of personal information, is enormous and easily accessible, as compared to equivalent access to such information in real space; and fourth, altogether the quantity of personal information on people over the Internet might be substantial. These characteristics imply that much power has been granted to individuals in their attempts to present themselves over the Internet, and by the same token, fellow Internet users have been granted even more power in their seeking of information on the personal identity of others.

Chapter 7 moved us to the third and last part of the book, and it was devoted to the Internet as an action space for individuals in their routine daily activities. Several leading online action spaces were presented: home-based work; online shopping; e-government; online banking; travel online; e-learning and e-health. The review of these online activities points to complementarity as the prevailing relationship, between real and virtual action spaces, rather than to full substitution of the real by the virtual. Thus, for each of the reviewed activities there is some advantage in doing business online, side-by-side with still some or much need for real space action, as well.

The performance of daily activities online has increased through the growing use of mobile access to the Internet, mainly through the easy to use smartphone versions of browsers and websites (or applications), which have made it possible for users to consume services without limitations of time and location in their routine daily life. This spatio-temporal flexibility amplifies even further the blurring of separation between the temporal and locational dimensions of daily activities. However, gaining experience in the use of smartphone applications, may potentially, at least, change the balance between services offered through facilities located in real space within easy physical reach by users, on the one hand, and virtual services provided through the Internet, or services located anywhere in real space and reached through the Internet, on the other. Such possible changes would potentially be in favor of the virtual and at the expense of the real, but would not necessarily lead to a full nullification of commercial service provision in real space.

The relatively veteran virtual action space of work has turned out to still be the most modest one, as far as its full adoption by individuals is concerned. The trends in both the US and the EU point to constant growth in the number and percentage of those working from home, including both those who work only from home and those who work primarily from home, but this pattern of location for work is still less preferred, so that it still consists of single-figure percentages of the workforce. In other words, only few people, in countries that lead in the personal adoption of Internet connectivity, are ready to use a virtual action space exclusively for their work, and the vast majority of workers still prefer some kind of a mix between physical and real action spaces for their work. Face-to-face work and work supervision are still highly valued by employees and employers, respectively.

The data for online shopping presented percentages of the population doing some but not all of their shopping online. The overall picture of the seemingly rather modest percentages of the population performing online shopping, mainly in the EU, seems to suggest that the geographical location and the preference for traditional stores located in real space are still valued by customers and might well continue to be of importance to customers in the future. This tendency for shopping in real action space, pertaining to shopping by individuals, still prevails because of shoppers' attitudes, reflected in habits, trust and uncertainties, which are main obstacles for a wider adoption of online shopping. These factors are accompanied by shoppers' preferences for the touching and trying of merchandise, and the perception of real space shopping as constituting also a kind of entertainment or amusement.

E-government has become widely offered and adopted in developed countries but it still awaits even further growth, which may eventually come about with additional growth in the adoption of the Internet by citizens, as a result of future declining prices of IT hardware and services, coupled with growing ease of their use, thus making it more affordable to the elderly and the poor. Lack of trust of governmental e-services by citizens might also be an obstacle for wide e-government adoption, and lack of competition in the provision of e-government services might also slow down the pace of development of easier to use online governmental services.

Another and rather commercial activity, which requires much trust, is banking. It seems that the use of virtual action space for banking involves not only a reasonable capability of subscribers to use banking websites and to understand well the basics of financial activities, but it implies also customers' trust in online banking services, and much more so than for real space banking services, mainly because of customers' apprehension concerning the quality of security measures required for online banking transactions.

The travel industry involves payment for purchased services through the online financial transactions, similarly to those developed for shopping online. However, the information on travel arrangements sought for by customers prior to such transactions is much more extensive than for the shopping of material products. Such information is required on potentially visited places and attractions there, as well as on potential service suppliers, such as hotels, in addition to the rather complex information which is required by potential tourists on the travel itself, mainly on flights. Since the major feature of the Internet as an action space is that it is an information space, travel online has developed fast and may develop even further in the future, notably in the supply of touristic information.

Distance learning might well be the most complex action space in its attempt to sell knowledge, as compared with the sale and provision of products, services and information, offered by the other virtual action spaces. Distance learning is provided not only for the provision of academic knowledge, but for the provision of professional knowledge as well. It seems that here too the preferred mode for both students and institutions of higher learning is some mix of face-to-face courses and video ones, whether pre-recorded or live ones.

E-health includes a wide variety of virtual activities consisting of management, consulting and treatment of one's health. The management of people's health life, as well as their consulting of health-related Web information have become routine activities in developed countries. The management component of e-health consists of an action space, similarly to the ones that emerged for other daily "e-" activities, whereas Web searches for health-related information are part of the wider tendency to satiate one's prudential curiosity through the Web. Telemedicine, or distance medical treatment, is relevant mostly to rural and peripheral areas in developed countries, and to developing countries at large. However, despite the availability of telemedicine technologies, there still exists a digital gap in this regard between high-income countries and other ones.

Chapter 8 focused on the seemingly most popular online action space, namely social networking. The Internet permits individuals to maintain wide geographical and social action spaces, and it further facilitates the management of wide social contacts as well as the management of the contents shared among them. These traits of online social networks stem from the very constitution of the Internet as an informational action space, which is, obviously, not the case for real action space per se. A major feature of Internet communications for online social networking is the joint locational and temporal flexibility, in that it does not require the synchronous attendance of the communicating parties.

Social networking, as part and parcel of the networked information economy, may be viewed as individual actions, coordinated with those of other people's individual actions, aided by the distributed mechanisms provided by information technology through proper platforms, and thus jointly bringing about social ties of numerous types and forms. The networked individuals present a new sense of personal, albeit connected, autonomy, expressed also through increased personal mobilities.

The very nature of social relations that develop online through social networks, as well as their potential impact on social relations in real space, has been widely debated in recent years. This debate has taken place in parallel with the massive adoption of social networking by Internet users worldwide, so that societies all over the globe are still in search of balances between real and virtual social action spaces and related social relations. It is assumed, though, that repeated virtual communications among parties may frequently lead to eventual face-to-face meetings of these parties.

We may quite safely assume that by 2013 some three-quarters of the Internet users worldwide were engaged in social networking over the net. Thus, the barrier-free, constraint-free (in most countries) and cost-free action space for virtual social networking has exhibited a tremendously fast adoption. However, the very use of social networking over the Internet does not automatically imply globally stretched networks attended by interested subscribers, as teens, for instance, use networking mainly for maintaining contacts with their classmates. Also, in the development of the Internet into a second action space for individuals, social networking has turned out to be by far, and at the global scale, the most popular activity taking place over Internet-based virtual action spaces. In developed countries, online social networking is as popular as other activities taking place online, whereas among Internet users in developing countries (a minority of the total population in most cases), it is even more popular than in developed ones, and this trend is accentuated even further by the weak uses of the Internet for other activities, mostly those of a more economic nature.

In Chapter 9 we turned our attention to darker Internet activities, which may be performed by governments, groups and by individuals. The joint effect of darker activities by all of these potential actors imply the limiting of free, full and safe access and action over the Internet by innocent users worldwide. This is true for the case of identity theft, the leading cybercrime, which attempts to benefit some criminal actors at the expense of the damaging of numerous innocent others. This is also true for other dark sides of the Internet, consisting of cyberobstruction measures which are instituted by governments in an attempt to avoid actions by others, mostly in form of censorship imposed by governments in order to avoid or in order to direct the Internet actions by their citizens, notably with regard to social networking. Similarly, this is further true also for hacking, whether performed by individuals, groups or governments.

Questions of potential limits to personal actions coupled with potential limits of governmental actions have accompanied the long development of societies in real space, and they have been imported into virtual space as well, with an obvious special twist for the informational and the rather global virtual space. The debate and the fight between the acceptable and the non-acceptable, between the permitted and the forbidden, notably with regard to pornography and online gambling, will accompany the future of Internet action spaces, presenting differences among countries on both political and cultural bases.

## Real and virtual action spaces

We noted along Parts II–III of the book four types of individual activities which are performed in virtual action spaces, fully or partially in equivalence to human activities traditionally taking place in real space: personal (basic needs, curiosity and personal identity), economic (work, shopping, banking, travel, etc.), social (networking) and dark (surveillance and identity theft). All of these activities are hosted on and are functioning through the same wide Internet platforms: the Web and email. Of these, the social action space has been found to be the most popular one worldwide. Also of these, some activities have been relatively restricted in real space and have been amplified in the virtual one, such as the satiation of curiosity, the presentation of personal identity, and again social networking. Economic activities which have become facilitated for online performance, such as shopping, banking, work, study, government services and travel services, are growingly being adopted by Internet users, mostly in complementarity with real world activities, but still online facilities and providers may compete in some way with those offered by real world ones. The relationships between real and virtual spaces seem to be more complex as far as personal services are concerned, mainly for human basic needs, since the relationships between these two spaces can range from complementarity through competition to substitution.

As we noted in the first part of the book most of the literature on the Internet focused on the relations between real and virtual spaces from the perspective of their constitution as spaces and systems of operation, or from the supply side of Internet use. We preferred to follow, for the second and third parts of the book, the approach that was preferred by Loo (2012) and Warf (2013), namely looking at the demand side of Internet users. Thus, we have attempted to assess the relations between the two spaces from the perspectives of human action and individual actors, or from the perspective of demand rather than that of supply.

Old geographical wisdom has claimed that individuals assess their action spaces via the place utility that they attribute to each of their action spaces (see, e.g., Jakle *et al.* 1976). Such place utilities refer to the levels of desirability, usefulness and satisfaction which individuals assign to each of their action spaces, thus yielding personal locational choices for their actions. Though we have focused in this book on virtual action spaces, it seems to be still too early to assess such action spaces vis-à-vis their place utilities for individual users. Whereas the choice for real action spaces for daily activities has been the focus of numerous areas of specialty within geography, notably in urban and economic geography, the establishment of clear and widely relevant rules for the assessment of place utilities for virtual action spaces seems to be still unfeasible as of yet. Virtual action spaces are not yet mature for their examination from the perspective of place utility, because they are still in the phase of their very making, and this is so for all of their major dimensions: website structures and tools for their operation by Internet users; devices and channels for access of the Internet; and above all the wide adoption of virtual action spaces for routine daily activities which is still just emerging.

Järv *et al.* (2014) have recently studied full-year data of mobile phone calls made by a sample of Tallinn residents throughout the year 2009, as subscribers of the largest mobile phone company in Estonia. They explored callers' activity spaces in real space, thus finding that on the average, callers visited some 35 physical sites per month, with ten of these sites being the most frequently visited ones. This type of study may serve as a benchmark for future similar studies attempting to discover similarly the number of websites visited by Internet users. Such studies will become feasible once the use of virtual websites and the place utilities attributed to them will become stabilized, probably in the upcoming years (for a discussion of the recent still unsettled conditions see Schwanen and Kwan 2008). It would be of interest and importance by then to compare the average number of websites visited per month by individuals, and those that are visited most frequently, to the number of physical sites visited in parallel. Furthermore, it would be of interest to compare individuals' double activity spaces, attempting to study the patterns of relationships among the lists of these sites for potential complementarity, competition or substitution among them.

The specific balances between online and offline human actions are not only system-dependent, namely depending on domestic infrastructures, cultures and habits, but these relations may turn out to be also user specific, as people may differ in their preferences for actions in real and virtual spaces. Furthermore, the relations between the two spaces need not be fixed, as they may change by circumstances from time to time. One of the major features of the emergence of online action space for individuals is that it provides for versatility, flexibility and convenience of operations, thus turning actors into more sophisticated human agents in all spheres of consumption and networking, whether performed in real or virtual action spaces. Still, as we noted, for most people there is a more fixed balance between actions performed in real and virtual spaces as far as the location of their work is concerned, but they prefer much more flexibility as far as shopping and shopping-related activities, such as price comparisons, are concerned. There is, though, one specific activity which has moved almost completely into virtual space, namely the search for information for almost all purposes, and relating to both epistemic and prudential curiosities, with the latter sometimes leading to "coveillance."

Under the contemporary circumstances of human action in double space, could we relate to contemporary human action as splintered between real and virtual action spaces? The notion of splintering has been proposed several times in the past, for several, though close to each other, geographical contexts. Thus, Graham and Marvin (2001) advanced the notion of splintering urbanism, focusing on cities as entities that possibly undergo some splintering processes. By the same token, Malecki and Moriset (2008), in their study of the digital economy, presented the possible splintering of the real economic space through communications. The difference between splintering cities and splintering economic spaces, on the one hand, and that of human action, on the other, is that the splintering of cities and economic spaces refers to the splintering of spatial units, whereas action does not constitute a spatial unit, and its splintering between two spaces implies the splintering of human action, brought about by the introduction of a new action space, the virtual one. Real social and economic spaces have not necessarily splintered because of the introduction and adoption of the novel virtual space, but human action, which is now maneuvered by actors between the two spaces, implies now the emergence of splintered action and splintered actors between two spaces. Each of these two spaces per se has not splintered because of the splintering of human action between the two of them.

## Time and virtual action space

We noted already in Chapter 1 the assessment of the pre World War II era in the US, and the following post-war era in Europe as ages dominated by an accent on speed, and this was based mainly on the wide adoption of personal mobility media of the time, notably the automobile for corporeal mobility and the telephone for virtual one (see, e.g., Virilio 1983; Freund and Martin 1993). The automobile made it possible to drastically reduce the friction of distance, while permitting face-to-face contacts, whereas the telephone nullified the friction of distance, sometimes at high costs at the time, and permitting a rather limited transmission of information, but requiring strict co-presence for conversations without face-to-face contacts (Kellerman 2006, 2012a). We may, thus, refer to the telephone and car era as the *age of speed*.

The later introduction of the Internet, followed by the introduction of the smartphone, have even further accentuated the nature of our age as based on speed, since these two tools have facilitated the transmission of all types of information and communications, both instantly and in delayed patterns. Hence, in addition to the past drastic diminishing or even nullification of distance, it has become now possible also for the time of reaction or response to messages to be too drastically diminished or even nullified. This cut in the time of communications has brought about social expectations for faster paces of communications. Thus, the expectation of email senders for receipt of responses to their email messages has been measured by hours and days, dropping to minutes and hours for SMSs, and going even further down to expectations for instant responses for chat messages, such as those transmitted through chat platforms, for instance WhatsApp and Viber.

The expectation of Internet users for their online performances of services, purchases, etc. is for 24/7 automatic or manual service availability, and for rather instant reaction and completion of service sessions. The frequent and even addicted use of network platforms by networked subscribers is yet another reflection of this diminishing of the dimension of time. Thus, the emergence of all types of online action spaces has been as significant for our temporal operation as it has been for our spatial one. We may refer, therefore, to our age as typified by *presentism*, though in a different sense than its normal meaning as a present interpretation of the past (e.g., Merriam-Webster 2013). Presentism in our age seems to refer to a contemporary expectation to have all actions take place in the immediate present and involving just minimal time duration (Rushkoff 2013). Moreover, the conception and experiencing of time in the information era has changed: "Time in the digital era is no longer linear but disembodied and associative" (Rushkoff 2013: 85), and time itself has been assumed to have changed its nature: "time itself becomes just another form of information – another commodity – to be processed" (Rushkoff 2013: 86). This contemporary perception of time and its use has further brought about its compression (see, e.g., Harvey 1989).

# Future virtual action spaces

In closing this book some thoughts on potential future trends rise in one's mind, dealing with several questions: will people live easily, in the long run, in the two spaces, the real and the virtual ones? What is the potential future of these two spaces? What is the technological horizon for future developments of virtual space?

It is most difficult to assess the future of virtual action space, mainly for two reasons. First is its dependence on future technological innovations. Since the Internet and smartphones are in the current forefront of continuous hi-tech efforts and developments, it is difficult to foresee the various dimensions of future virtual action spaces. Second, it is further difficult to speculate on the rate of adoption of new waves of information technologies. A major possible future development is the Internet of Everything, which will attempt to connect household and other devices directly to the Internet, thus potentially creating completely new action spaces (see Haynes and Campbell 2013).

So far, the emergence of virtual action space and its adoption has attested to the ability and willingness of people to act in parallel and in an integrative way within the two spaces, the real and virtual. However, as experience has shown, not every future technological innovation may be adopted, and in other cases, it sometimes may take several decades and numerous technological generations until a device or software gets adopted, as happened, for instance, in the 1970s with the then popular apathy to the introduction of videophones and hand-held mini-computers, two technologies which have become widely adopted 30 years later on. However, even if we assume a currently unrealistic assumption of an upcoming saturation in Internet-based technological innovations, the continuing adoption of currently available online actions may potentially bring about more radical changes in the urban real-space service economy, side-by-side with possible social changes in real space to be brought about by online social relations.

# Appendix

Percentage of individuals using the Internet for selected countries 2000–2012

| Country | 2000 | 2001 | 2002 | 2003 | 2004 | 2005 | 2006 | 2007 | 2008 | 2009 | 2010 | 2011 | 2012 |
|---|---|---|---|---|---|---|---|---|---|---|---|---|---|
| Argentina | 7.04 | 9.78 | 10.88 | 11.91 | 16.04 | 17.72 | 20.93 | 25.95 | 28.11 | 34.00 | 45.00 | 51.00 | 55.80 |
| Armenia | 1.30 | 1.63 | 1.96 | 4.58 | 4.90 | 5.25 | 5.63 | 6.02 | 6.21 | 15.30 | 25.00 | 32.00 | 39.16 |
| Australia | 46.76 | 52.69 | N/A | N/A | N/A | 63.00 | 66.00 | 69.45 | 71.67 | 74.25 | 76.00 | 79.50 | 82.35 |
| Austria | 33.73 | 39.19 | 36.56 | 42.70 | 54.28 | 58.00 | 63.60 | 69.37 | 72.87 | 73.45 | 75.17 | 79.80 | 81.00 |
| Bangladesh | 0.07 | 0.13 | 0.14 | 0.16 | 0.20 | 0.24 | 1.00 | 1.80 | 2.50 | 3.10 | 3.70 | 5.00 | 6.30 |
| Belgium | 29.43 | 31.29 | 46.33 | 49.97 | 53.86 | 55.82 | 59.72 | 64.44 | 66.00 | 70.00 | 75.00 | 78.00 | 82.00 |
| Botswana | 2.90 | 3.43 | 3.39 | 3.35 | 3.30 | 3.26 | 4.29 | 5.28 | 6.25 | 6.15 | 6.00 | 8.00 | 11.50 |
| Brazil | 2.87 | 4.53 | 9.15 | 13.21 | 19.07 | 21.02 | 28.18 | 30.88 | 33.83 | 39.22 | 40.65 | 45.00 | 49.85 |
| Cambodia | 0.05 | 0.08 | 0.23 | 0.26 | 0.30 | 0.32 | 0.47 | 0.49 | 0.51 | 0.53 | 1.26 | 3.10 | 4.94 |
| Canada | 51.30 | 60.20 | 61.59 | 64.20 | 65.96 | 71.66 | 72.40 | 73.20 | 76.70 | 80.30 | 80.30 | 83.00 | 86.77 |
| Chad | 0.04 | 0.05 | 0.17 | 0.32 | 0.36 | 0.40 | 0.58 | 0.85 | 1.19 | 1.50 | 1.70 | 1.90 | 2.10 |
| Chile | 16.60 | 19.10 | 22.10 | 25.47 | 28.18 | 31.18 | 34.50 | 35.90 | 37.30 | 41.56 | 45.00 | 52.25 | 61.42 |
| China | 1.78 | 2.64 | 4.60 | 6.20 | 7.30 | 8.52 | 10.52 | 16.00 | 22.60 | 28.90 | 34.30 | 38.30 | 42.30 |
| Colombia | 2.21 | 2.85 | 4.60 | 7.39 | 9.12 | 11.01 | 15.34 | 21.80 | 25.60 | 30.00 | 36.50 | 40.40 | 48.98 |
| Cuba | 0.54 | 1.08 | 3.77 | 5.24 | 8.41 | 9.74 | 11.16 | 11.69 | 12.94 | 14.33 | 15.90 | 23.23 | 25.64 |
| Egypt | 0.64 | 0.84 | 2.72 | 4.04 | 11.92 | 12.75 | 13.66 | 16.03 | 18.01 | 25.69 | 31.42 | 39.83 | 44.07 |
| Ethiopia | 0.02 | 0.04 | 0.07 | 0.11 | 0.16 | 0.22 | 0.31 | 0.37 | 0.45 | 0.54 | 0.75 | 1.10 | 1.48 |
| Finland | 37.25 | 43.11 | 62.43 | 69.22 | 72.39 | 74.48 | 79.66 | 80.78 | 83.67 | 82.49 | 86.89 | 89.37 | 91.00 |
| France | 14.31 | 26.33 | 30.18 | 36.14 | 39.15 | 42.87 | 46.87 | 66.09 | 70.68 | 71.58 | 80.10 | 79.58 | 83.00 |
| Germany | 30.22 | 31.65 | 48.82 | 55.90 | 64.73 | 68.71 | 72.16 | 75.16 | 78.00 | 79.00 | 82.00 | 83.00 | 84.00 |

| | | | | | | | | | | | | | |
|---|---|---|---|---|---|---|---|---|---|---|---|---|---|
| Ghana | 0.15 | 0.20 | 0.83 | 1.19 | 1.72 | 1.83 | 2.72 | 3.85 | 4.27 | 5.44 | 7.80 | 14.11 | 17.11 |
| India | 0.53 | 0.66 | 1.54 | 1.69 | 1.98 | 2.39 | 2.81 | 3.95 | 4.38 | 5.12 | 7.50 | 10.07 | 12.58 |
| Indonesia | 0.93 | 2.02 | 2.13 | 2.39 | 2.60 | 3.60 | 4.76 | 5.79 | 7.92 | 6.92 | 10.92 | 12.28 | 15.36 |
| Iran | 0.93 | 1.48 | 4.63 | 6.93 | 7.49 | 8.10 | 8.76 | 9.47 | 10.24 | 11.07 | 14.70 | 21.00 | 26.00 |
| Italy | 23.11 | 27.22 | 28.04 | 29.04 | 33.24 | 35.00 | 37.99 | 40.79 | 44.53 | 48.83 | 53.68 | 56.80 | 58.00 |
| Japan | 29.99 | 38.53 | 46.59 | 48.44 | 62.39 | 66.92 | 68.69 | 74.30 | 75.40 | 78.00 | 78.21 | 79.05 | 79.05 |
| Mexico | 5.08 | 7.04 | 11.90 | 12.90 | 14.10 | 17.21 | 19.52 | 20.81 | 21.71 | 26.34 | 31.05 | 34.96 | 38.42 |
| Morocco | 0.69 | 1.37 | 2.37 | 3.35 | 11.61 | 15.08 | 19.77 | 21.50 | 33.10 | 41.30 | 52.00 | 53.00 | 55.00 |
| Netherlands | 43.98 | 49.37 | 61.29 | 64.35 | 68.52 | 81.00 | 83.70 | 85.82 | 87.42 | 89.63 | 90.72 | 92.30 | 93.00 |
| Nigeria | 0.06 | 0.09 | 0.32 | 0.56 | 1.29 | 3.55 | 5.55 | 6.77 | 15.86 | 20.00 | 24.00 | 28.43 | 32.88 |
| Pakistan | N/A | 1.32 | 2.58 | 5.04 | 6.16 | 6.33 | 6.50 | 6.80 | 7.00 | 7.50 | 8.00 | 9.00 | 9.96 |
| Poland | 7.29 | 9.90 | 21.15 | 24.87 | 32.53 | 38.81 | 44.58 | 48.60 | 53.13 | 58.97 | 62.32 | 64.88 | 65.00 |
| Russia | 1.98 | 2.94 | 4.13 | 8.30 | 12.86 | 15.23 | 18.02 | 24.66 | 26.83 | 29.00 | 43.00 | 49.00 | 53.27 |
| Saudi Arabia | 2.21 | 4.68 | 6.38 | 8.00 | 10.23 | 12.71 | 19.46 | 30.00 | 36.00 | 38.00 | 41.00 | 47.50 | 54.00 |
| South Africa | 5.35 | 6.35 | 6.71 | 7.01 | 8.43 | 7.49 | 7.61 | 8.07 | 8.43 | 10.00 | 24.00 | 33.97 | 41.00 |
| Sweden | 45.69 | 51.77 | 70.57 | 79.13 | 83.89 | 84.83 | 87.76 | 82.01 | 90.00 | 91.00 | 90.00 | 94.00 | 94.00 |
| Switzerland | 47.10 | 55.10 | 61.40 | 65.10 | 67.80 | 70.10 | 75.70 | 77.20 | 79.20 | 81.30 | 83.90 | 85.20 | 85.20 |
| Turkey | 3.76 | 5.19 | 11.38 | 12.33 | 14.58 | 15.46 | 18.24 | 28.63 | 34.37 | 36.40 | 39.82 | 43.07 | 45.13 |
| UK | 26.82 | 33.48 | 56.48 | 64.82 | 65.61 | 70.00 | 68.82 | 75.09 | 78.39 | 83.56 | 85.00 | 86.84 | 87.02 |
| US | 43.08 | 49.08 | 58.79 | 61.70 | 64.76 | 67.97 | 68.93 | 75.00 | 74.00 | 71.00 | 74.00 | 77.86 | 81.03 |
| Venezuela | 3.36 | 4.64 | 4.91 | 7.50 | 8.40 | 12.55 | 15.22 | 20.83 | 25.88 | 32.70 | 37.37 | 40.22 | 44.05 |

Data source: ITU 2013. See source for country specific sources.

# References

About.com (2013) "2013 Facebook global user statistics." Online. Available at: http://womeninbusiness.about.com/od/facebook/a/2013-Facebook-User-Statistics.htm (accessed October 21, 2013).

Abt, H.A. (2007) "The publication rate of scientific papers depends only on the number of scientists," *Scientometrics*, 73: 281–8.

Acoca, B. (2008) "Online identity theft," *The OECD Observer*, 268: 12–13.

Adams, P.C. (1995) "A reconsideration of personal boundaries in space-time," *Annals of the Association of American Geographers*, 85: 267–85.

Adams, P.C. and Ghose, R. (2003) "India.com: The construction of a space between," *Progress in Human Geography*, 27: 414–37.

Adams, P.C. and Skop, E. (2008) "The gendering of Asian Indian transnationalism on the Internet," *Journal of Cultural Geography*, 25: 115–36.

Adelfer, C. (1972) *Existence, Relatedness and Growth: Human Needs in Organizational Setting*, New York: Free Press.

Adelfer, C. (1989) "Theories reflecting my personal experience and life development," *Journal of Applied Behavioral Science*, 25: 351–65.

Agar, J. (2003) *Constant Touch: A Global History of the Mobile Phone*, Cambridge: Revolutions in Science.

Ahas, R., Silm, S., Järv, O. Saluveer, E. and Tiru, M. (2010) "Using mobile positioning data to model locations meaningful to users of mobile phones," *Journal Of Urban Technology*, 17: 3–27.

Allot Communications (2013) "Allot mobile trends report 02/2013." Online. Available at: www.allot.com/Press_Releases.html (accessed May 27, 2013).

Aguiléra, A., Guillot, C. and Rallet, A. (2012) "Mobile ICTs and physical mobility: Review and research agenda," *Transportation Research A*, 46: 664–72.

Allen, J. (1999) "Worlds within cities," in D. Massey, J. Allen and S. Pile (eds.) *City Worlds*, London: Routledge, 53–98.

American Heritage Bank (2013) "Online banking statistics." Online. Available at: www/bankadviser.com/americanheritagebank/e_article002635091.cfm?x+b11,0,w (accessed June 2, 2013).

Aoyama, Y. (2003) "Sociospatial dimensions of technology adoption: Recent M-commerce and E-commerce developments," *Environment and Planning A*, 35: 1201–21.

Aristotle, *Metaphysics*, W.D. Ross (trans.) (1953). Online. Available at: http://ebooks.adelaide.edu.au/ (accessed April 1, 2012).

Aristotle, *Politics*, B. Jowett (trans.). Online. Available at: http://classics.mit.edu/Aristotle/politics.html (accessed June 19, 2013).

Ariunaa, L. (2006) "Mongolia: Mobilizing communities for participation in e-government initiatives for the poor and marginalized," *Regional Development Dialogue*, 27: 140–51.

Arminen, I. (2007) "Review essay: Mobile communication society?," *Acta Sociologica*, 50: 431–7.

Asheim, B., Coenen, L. and Vang, J. (2007) "Face-to-face, buzz, and knowledge bases: Sociospatial implications for learning, innovation and innovation policy," *Environment and Planning C: Government and Policy*, 25: 655–70.

Atkinson, R. (1998) "Technological change and cities," *Cityscape: A Journal of Policy Development and Research*, 3: 129–71.

Aud, S., Hussar, W., Kena, G., Bianco, K., Frohlich, L., Kemp, J. and Tahan, K. (2011) *The Condition of Education 2011* (NCES 2011–33). US Department of Education, National Center for Education Statistics. Washington, DC: US Government Printing Office.

Augé, M. (2000) *Non-Places: Introduction to an Anthropology of Supermodernity*, J. Howe (trans.) London: Verso.

Augoyard, J.-F. and Torgue, H. (2005) *Sonic Experience*, A. McCartney and D. Paquette (trans.) Montreal: McGill-Queens University Press.

Australian Government Information Management Office (2011) "Interacting with government: Australians' use and satisfaction with e-government services." Online. Available at: www.finance.gov.au/publications/interacting-with-government-2011/docs/interacting-with-government-2011.pdf (accessed June 2, 2012).

Barthes, R. (1972) *Mythologies*, A. Lavers (trans.) London: Jonathan Cape.

Basu, P. and Chakraborty, J. (2011) "New technologies, old divides: linking Internet access to social and locational characteristics of US farms," *GeoJournal*, 76: 469–81.

Bates, P. and Huws, U. (2002) "Modelling ework in Europe: Estimates, Models and Forecasts from the EMERGENCE Project." Falmer, Brighton: Institute for Employment Studies. Online. Available at: www.employment-studies.co.uk/pdflibrary/388.pdf (accessed May 29, 2013).

Bathelt, H. and Schuldt, N. (2008) "Temporary face-to-face-contact and the ecologies of global and virtual buzz," *SPACES online 6 2008–04*. Online. Available at: www.spaces-online.uni-hd.de/ (accessed June 27, 2013).

Batty, M. (1997) "Virtual geography," *Futures*, 29: 337–52.

Benedikt, M. (1991) "Cyberspace: Some proposals," in M. Benedikt (ed.) *Cyberspace: First Steps*, Cambridge, MA: MIT Press, 119–224.

Benkler, Y. (2006) *The Wealth of Networks: How Social Production Transforms Markets and Freedom*, New Haven: Yale University Press.

Ben-Ze'ev, A. (2004) *Love Online: Emotions on the Internet*, Cambridge: Cambridge University Press.

Betz, D. (2012) "Cyberpower in strategic affairs: Neither unthinkable nor blessed," *The Journal of Strategic Studies*, 35: 689–711.

Bialski, P. (2012) *Becoming Intimately Mobile*, Frankfurt: Peter Lang.

Biegel, S. (2001) *Beyond Our Control?: Confronting the Limits of Our Legal System in the Age of Cyberspace*, Cambridge, MA: MIT Press.

Blainey, A. (1966) *Tyranny of Distance: How Distance Shaped Australia's History*, Melbourne: Macmillan.

Blum, A. (2012) *Tubes: A Journey to the Center of the Internet*, New York: Ecco.

Boden, D. and Molotch, H.L. (1994) "The compulsion of proximity," in R. Friedland and D. Boden (eds.) *NowHere Space, Time and Modernity*, Berkeley: University of California Press, 257–86.

Bolter, J.D. and Grusin, R. (1999) *Remediation: Understanding New Media*, Cambridge, MA: MIT Press.

Boyd, D.M. and Ellison, N.B. (2007) "Social networking sites: Definition, history, and scholarship," *Journal of Computer-mediated Communication*, 13. Online. Available at: http://jcmc.indiana.edu/vol. 13/issue1/boyd.ellison.html (accessed June 20, 2013).

Brain Track (2013) "Women lag men by 25% in Internet adoption." Online. Available at: www.braintrack.com/blog/2013/02/women-lag-men-by-25-in-internet-adoption/ (accessed October 21, 2013).

Breen, G.-M. and Matusitz, J. (2010) "An evolutionary examination of E-health: A health and computer-mediated communication perspective," *Social Work in Public Health*, 25: 59–71.

Brenner, J. (2012) "Pew Internet: Teens." Online. Available at: http://pewinternet.org/Commentary/2012/April/Pew-Internet-Teens.aspx (accessed October 22, 2013).

Breton, G. and Lambert, M. (eds.) (2003) *Universities and Globalization: Private Linkages, Public Trust*, Paris: UNESCO and Université Laval.

Brown, S. (2005) "Travelling with a purpose: Understanding the motives and benefits of volunteer vacationers," *Current Issues in Tourism*, 8: 479–96.

Brunn, S.D. and Leinbach, T.R. (eds.) (1991) *Collapsing Space and Time: Geographic Aspects of Communication and Information*, New York: Harper Collins Academic.

Bruns, A. (2008) *Blogs, Wikipedia, Second Life, and Beyond: From Production to Produsage*, New York: Peter Lang.

Cairncross, F. (1997) *The Death of Distance: How the Communications Revolution Will Change Our Lives*, Boston: Harvard Business School Press.

Capineri, C. and Leinbach, T.R. (2004) "Globalization, e-economy and trade," *Transport Reviews*, 24: 645–63.

Castells, M. (1989) *The Informational City: Information, Technology, Economic Restructuring and the Urban-Regional Process*, Oxford: Blackwell.

Castells, M. (2000) *The Rise of the Network Society*, 2nd ed. Oxford: Blackwell.

Castells, M. (2001) *The Internet Galaxy: Reflections on the Internet, Business, and Society*, New York: Oxford University Press.

Castells, M. (2009) *Communication Power*, Oxford: Oxford University Press.

Castells, M., Fernánddez-Ardèvol, M., Qiu, J.L. and Sey, A. (2007) *Mobile Communication and Society: A Global Perspective*, Cambridge, MA: MIT Press.

Cadwick, A. and May, C. (2003) "Interaction between states and citizens in the age of the Internet: 'E-government' in the United States, Britain and the European Union," *Governance*, 16: 271–300.

ChartsBin (2010) "Legal status of child pornography by country." Online. Available at: http://chartsbin.com/view/q4y (accessed July 10, 2013).

Checkfacebook (2013) "Analyze and enhance Facebook performance." Online. Available at: www.checkfacebook.com/ (accessed April 19, 2013).

CIA (Central Intelligence Agency) (2013) "The world factbook," Online. Available at: www.cia.gov/library/publications/the-world-factbook/rankorder/2153rank.html (accessed April 14, 2013).

Coker, B.L.S. (2011) "Freedom to surf: The positive effects of workplace Internet leisure browsing," *New Technology, Work and Employment*, 26: 238–47.

Comer, J. C. and Wikle, T. A. (2008) "Worldwide diffusion of the cellular telephone, 1995–2005," *The Professional Geographer*, 60: 252–69.

Commission of the European Communities (2009) "Report on cross-border e-commerce in the EU," Online. Available at: http://ec.europa.eu/consumers/strategy/docs/com_staff_wp2009_en.pdf (accessed May 30, 2013).

ComScore (2012) "ComScore: 4 out of 5 smartphone owners use device to shop: Amazon is the most popular mobile retailer." Online. Available at: http://techcrunch.com/2012/09/19/comscore-4-out-of-5-smartphone-owners-use-device-to-shop-amazon-most. (Accessed May 28, 2013).

Connor, S. (1997) "The modern auditory I," in R. Porter (ed.) *Rewriting the Self: Histories from the Renaissance to the Presence*, London and New York: Routledge, 203–23.

Cosgrove, D. (1984) *Social Formation and Symbolic Landscape*, New Jersey: Barnes & Noble.

Couclelis, H. (1998) "Worlds of Information: The geographic metaphor in the visualization of complex information," *Cartography and Geographic Information Systems*, 25: 209–20.

Couclelis, H. (2004) "Pizza over the Internet: E-commerce, the fragmentation of activity and the tyranny of the region," *Entrepreneurship and Regional Development*, 16: 41–54.

Coursera (2013) Online. Available at: www.coursera.org/ (accessed April 30, 2013).

Cowen, T. (2011) *The Great Stagnation: How America Ate All the Low-hanging Fruit of Modern History, Got Sick, and Will (Eventually) Feel Better*, New York: Penguin Group.

Crampton, J.W. (2007) "The biopolitical justification for geosurveillance," *Geographical Review*, 97: 389–403.

Crang, M., Crang, P. and May, J. (eds.) (1999) *Virtual Geographies: Bodies, Space and Relations*, London: Routledge.

Crandall, R. (1980) "Motivations for leisure," *Journal of Leisure Research*, 12: 45–54.

Crutcher, M. and Zook, M. (2009) "Placemarks and waterlines: Racialized cyberscapes in post-Katrina Google Earth," *Geoforum*, 40, 523–34.

Curry, M.R. (2000) "The power to be silent: Testimony, identity, and the place of place," *Historical Geography*, 28: 13–24.

Darnton, R. (2011) "Google's loss: The public's gain," *The New York Times Review of Books*, 18 April: 28.

David, P.A. (1990) "The dynamo and the computer: An historical perspective on the modern productivity paradox," *American Economic Review Papers and Proceedings*, 80: 355–61.

Davis, J. (2010) "Architecture of the personal interactive homepage: Constructing the self through MySpace," *New Media and Society*, 12: 1103–19.

De Blasio, G. (2008) "Urban-rural differences in Internet usage, e-commerce, and e-banking: Evidence from Italy," *Growth and Change*, 39: 341–67.

Dijst, M. (2004) "ICTs and accessibility: An action space perspective on the impact of new information and communications technologies," in M. Bethue, V. Himanen, A. Reggiani and L. Zamparini (eds.) *Transport Developments and Innovations in an Evolving World*, Berlin: Springer, 27–46.

Ding, W.W., Levin, S.G., Stephan, P.E. and Winkler, A.E. (2010) "The impact of information technology on academic scientists' productivity and collaboration patterns," *Management Science*, 56: 1415–38.

Digitalbuzz blog (2013) "Infographic: Social media statistics for 2012." Online. Available at: www.digitalbuzzblog.com/social-media-statistics-stats-2012-infographic/ (accessed July 5, 2013).

Dobson, J.E. and Fisher, P.F. (2007) "The Panopticon's changing geography," *The Geographical Review*, 97: 307–23.

Dodge, M. (2001) "Guest editorial," *Environment and Planning B: Planning and Design*, 28: 1–2.

Dodge, M. and Kitchin, R. (2001) *Mapping Cyberspace*, London: Routledge.

Dominick, J.R. (1999) "Who do you think you are? Personal home pages and self presentation on the World Wide Web," *Journalism and Mass Communication Quarterly*, 77: 646–58.

Dovbysh, O. (2013) "The peculiarities of using digital technologies in rural areas in Russia." Paper presented at the World Social Sciences Forum, Montreal.

Duggan, M. and Brenner, J. (2013) *The Demographics of Social Media Users – 2012*. PewInternet. Online. Available at: http://pewinternet.org/Reports/2013/Social-media-users.aspx (accessed November 27, 2013).

Duranton, G. (1999) "Distance, land, and proximity: Economic analysis and the evolution of cities," *Environment and Planning A*, 31: 2169–88.

Easton, S. (1994) *The Problem of Pornography: Regulation and the Right to Free Speech*, London and New York: Routledge.

*The Economist* (2012) "A sense of place," 27 October.

Entrikin, J.N. (1991) *The Betweenness of Place: Towards a Geography of Modernity*, Baltimore: The Johns Hopkins University Press.

Ettlinger, N. (2003) "Cultural economic geography and a relational and microspace approach to trusts, rationalities, networks and change in collaborative workplaces." *Journal of Economic Geography*, 3: 145–71.

European Commission (2002) "Qualitative study on cross border shopping in 28 European countries." Online. Available at: http://ec.europa.eu/consumers/topics/cross_border_shopping_en_pdf (accessed May 30, 2013).

European Commission (2006) "Consumer protection in the Internet market." Special Eurobarometer 252. Online. Available at: http://ec.europa.eu/public_opinion/archives/ebs/ebs252_en.pdf (accessed May 30, 2013).

European Commission (2010) "E-government statistics." Online. Available at: http://epp.eurosta.ec.europa.eu/statistics_explained/index.php/E-government_statistics (accessed May 31, 2013).

European Commission (2011) "Consumer attitudes towards cross-border trade and consumer protection: Analytical report." Flash Eurobarometer 299. Online. Available at: http://ec.europa.eu/public_opinion/flash/fl_299_wn.pdf (accessed May 30, 2013).

EU (European Union) (2006) *eUSER Population Survey 2005*. Online. Available at: www.euser-eu.org/eUSER_PopulationSurveyStatistics.asp (accessed June 5, 2013).

European Travel Commission New Media Travel Watch (2013) "Asia Pacific eCommerce." Online. Available at: www.newmediatrendwatch.com/regional-overview/90-asian?start=1 (accessed May 12, 2013).

Eurostat (2012a) "Internet use in households and by individuals in 2012." Online. Available at: http://epp.eurostat.ec.europa.eu/cache/ITY_OFFPUB/KS-SF-12-050/EN/KS-SF-12-050-EN.PDF (accessed May 30, 2013).

Eurostat (2012b) "Newsrelease: Internet access and use in 2012." Online. Available at: http://epp.eurostat.ec.europa.eu/cache/ITY_PUBLIC/4-18122012-AP/EN/4-18122012-AP-EN.PDF (accessed June 4, 2012).

Fabrikant, S.I. and Buttenfield, B.P. (2001) "Formalizing semantic spaces for information access," *Annals of the Association of American Geographers*, 91: 263–80.

Facebook (2012) "Statistics." Online. Available at: www.facebook.com/press/info.php?statistics (accessed April 1, 2012).

Farivar, C. (2011) *The Internet of Elsewhere: The Emergent Effects of a Wired World*, New Brunswick, NJ: Rutgers University Press.

FCC (Federal Communications Commission) (2009). "Statement of Chairman K. J. Martin." Online. Available at: http://hraunfoss.fcc.gov/edocs_public/attachmatch/DOC-280909A2.doc (accessed May 28, 2013).

Feldman, R.S. (2003) *Essentials of Understanding Psychology*, New York: McGraw Hill.

Felstead, A., Jewson, N. and Walters, S. (2005) *Changing Places of Work*, Basingstoke: Palgrave Macmillan.

Ferreira, A., Batey P., Brömmelstroet, M.T. and Bertolini, L. (2012) "Beyond the dilemma of mobility: Exploring new ways of matching intellectual and physical mobility," *Environment and Planning A*, 44: 688–704.

Forbes (2011) "How much of the Internet is actually for porn." Online. Available at: www.forbes.com/sites/julieruvolo/2011/09/07how-much-of-the-internet-is-actually-for-porn/ (accessed July 10, 2013).

Forbes (2013) "An overview why Microsoft's worth $42." Online. Available at: www.forbes.com/sites/greatspeculations/2013/01/09/an-overview-why-microsofts-worth-42/ (accessed July 10, 2013).

Foucault, M. (1972) *The Archeology of Knowledge*, London: Tavistock.

Fountain, J. (2000) "Constructing the information society: Women, information technology, and design," *Technology in Society*, 22: 45–62.

Fowler, H. (1965) *Curiosity and Exploratory Behavior*, New York: Macmillan.

Franck, G. (1998) *Ökonomie der Aufmerksamkei*, Munich: Hanser [German].

Freund, P. and Martin, G. (1993) *The Ecology of the Automobile*, Montreal: Black Rose Books.

Fuller, D. and Askins, K. (2010) "Public geographies II: Being organic," *Progress in Human Geography*, 34: 654–67.

Gade, D.W. (2011) *Curiosity, Inquiry, and the Geographical Imagination*, New York: Peter Lang.

Gartner (2013) "Gartner says worldwide mobile phone sales declined 1/7 percent in 2012." Online. Available at: www.gartner.com/newsroom/id/2335616 (accessed April 18, 2013).

Gergen, K.J. (1991) *The Saturated Self: Dilemmas of Identity in Contemporary Life*, New York: Basic Books.

Gergen, K.J. (1992) "The decline and fall of personality: How the fax, the phone, and the VCR are taking us beyond ourselves," *Psychology Today*, 25: 58–63.

Gibson, W. (1984) *Neuromancer*, London: Gollancz.

Giddens, A. (1981) *A Contemporary Critique of Historical Materialism*, Berkeley, CA: University of California Press.

Giddens, A. (1990) *The Consequences of Modernity*, Cambridge: Polity Press.

Gilbert, M.R., Masucci, M., Homko, C. and Bove, A.A. (2008) "Theorizing the digital divide: Information and communication technology use frameworks among poor women using a E-health system," *Geoforum*, 39: 912–25.

Gilbert, M.R. and Masucci, M. (2011) *Information and Communication Technology Geographies: Strategies for Bridging a Digital Divide*, Vancouver: Praxis (e) Press.

Gillespie, A. and Williams, H. (1988) "Telecommunications and the reconstruction of Regional Comparative Advantage," *Environment and Planning A*, 20: 1311–21.

Gillespie, A. and Robins, K. (1989) "Geographical inequalities: The spatial bias of the new communications technologies," *Journal of Communications*, 39: 7–18.

Glatzmeier, H. and Steinhardt, G. (2005) "Self presentations on personal homepages," in A. Sloane (ed.) *Home-Oriented Informatics and Telematics*, New York: Springer, 33–50.

Goffman, E. (1959) *The Presentation of Self in Everyday Life*, Garden City, NY: Double Day.

Goffman, E. (1961) *Asylums: Essays on the Social Situation of Mental Patients and Other Inmates*, Chicago: Aldine.

Gold, J.R. (1980) *An Introduction to Behavioural Geography*, Oxford: Oxford University Press.

Golden, W., Hughes, M. and Scott, M. (2003) "Implementing E-government in Ireland: A roadmap for success," *Journal of Electronic Commerce in Organizations*, 1: 17–33.

Golledge, R.G. and Stimson, R.J. (1997) *Spatial Behavior: A Geographic Perspective*, New York: Guilford Press.

Goodquotes.com (2013) "Laurence Sterne quotes." Online. Available at: www.goodquotes.com/quote/laurence-sterne/the-desire-of-knowledge-like-the-thirs (accessed April 30, 2013).

Gopal, S. (2007) "The evolving social geography of blogs," in H.J. Miller (ed.) *Societies and Cities in the Age of Instant Access*, Dordrecht: Springer, 275–93.

Gordon, E. and de Souza e Silva, A. (2011) *Net Locality: Why Location Matters in a Networked World*, Chichester: Wiley-Blackwell.

Gordon, R.J. (2010) "Revisiting U.S. productivity growth over the past century with a view of the future." NBER (National Bureau of Economic Research) Working Paper 15834. Online. Available at: www.nber.org/papers/w15834.pdf (accessed November 27, 2013).

Gorman, S.P. and Malecki, E.J. (2001) "Fixed and fluid: Stability and change in the geography of the Internet." Paper presented at the Annual Meeting of the Association of American Geographers, New York.

Graham, S. (1998) "The end of geography or the explosion of place? Conceptualizing space, place and information technology," *Progress in Human Geography*, 22: 165–85.

Graham, S. and Marvin, S. (1996) *Telecommunications and the City*, London and New York: Routledge.

Graham, S. and Marvin, S. (2001) *Splintering Urbanism: Networked Infrastructures, Technological Mobilities and the Urban Condition*, London: Routledge.

Greenwood, D.J. and Levin, M. (1998) *Introduction to Action Research: Social Research for Social Change*, Thousand Oaks, CA: Sage.

Gregory, D., Martin, M. and Smith, G. (1994) "Introduction: Human geography, social change and social science," in D. Gregory, R. Martin and G. Smith, (eds.) *Human Geography: Society, Space and Social Science*, London: Macmillan, 78–112.

Griffiths, F., Cave, J., Boardman, F., Justin, R., Pawlikowska, T., Ball, R., Clarke, A. and Cohen, A. (2012) "Social networks – the future for health care delivery," *Social Science and Medicine*, 75: 2233–41.

Groen, J., Smit, E. and Eijsvoogel, J. (eds.) (1990) *The Discipline of Curiosity: Science in the World*, Amsterdam: Elsevier.

Guo, Y. and Chen, P. (2011) "Digital divide and social cleavage: Case studies of ICT usage among peasants in contemporary China," *China Quarterly*, 207: 580–99.

Hägerstrand, T. (1970) "What about people in regional science?" *Papers and Proceedings of the Regional Science Association*, 24: 7–21.

Hägerstrand, T. (1973) "The domain of human geography," in R.J. Chorley (ed.) *Directions in Geography*, London: Methuen, 67–87.

Hägerstrand, T. (1975) "Space, time and human conditions," in A. Karlqvist, L. Lundqvist and F. Snickers (eds.) *Allocation of Urban Space*, Farnborough: Saxon House, 3–12.

Hägerstrand, T. (1992) "Mobility and transportation – are economics and technology the only limits?," *Facta and Futura*, 2: 35–38.

Häkli, J. and Paasi, A. (2003) "Geography, space and identity," in J. Öhman and K. Simonsen (eds.) *Voices from the North: New Trends in Nordic Human Geography*, Farnham, UK and Burlington, VT: Ashgate, 141–55.

Halbwachs, M. (1980) *The Collective Memory*, F.J. Dulles and V.Y. Ditter (trans.) New York: Harper and Row.

Halford, S. (2005) "Hybrid workspace: Re-spatialisations of work, organisation and management," *New Technology, Work and Employment*, 20: 19–33.

Hall, C.M. and Page, S.J. (2006) *The Geography of Tourism and Recreation: Environment, Place and Space*, 3rd ed. London and New York: Routledge.

Hall, S. (1996) "Who needs identity," in S. Hall and P. du Gay (eds.) *Questions of Cultural Identity*, London: Sage, 1–17.

Hanover Research (2011) "Trends in Global Distance Learning." Online. Available HTTP: www.hanoverresearch.com (accessed February 3, 2013).

Hargittai, E. (1999) "Weaving the Western web: Explaining differences in Internet connectivity among OECD countries," *Telecommunications Policy*, 23: 701–18.

Harvey, D. (1989) *The Coming of Postmodernity*. Oxford: Blackwell.

Hassa, S. (2012) "Projecting, exposing, revealing self in the digital world: Usernames as a social practice in a Moroccan chatroom," *Names*, 60: 201–9.

Haynes, P. and Campbell, T.A. (2013) "Hacking the Internet of Everything," *Scientific American*. Online. Available at: www.scientificamerican.com/article.cfm?id=hacking-internet-of-everything (accessed November 17, 2013).

Helminen, V. and Ristimäki, M. (2007) "Relationships between commuting distance, frequency and telework in Finland," *Journal of Transport Geography*, 15: 331–42.

Herring, S.C., Scheidt, L.A., Wright, E. and Bonus, S. (2005) "Weblogs as bridging genre," *Information, Technology and People*, 18: 142–71.

Hislop, D. and Axtell, C. (2007) "The neglect of spatial mobility in contemporary studies of work: The case of telework," *New Technology, Work and Employment*, 22: 34–51.

Holloway, S.I. and Valentine, G. (2001) "Placing cyberspace: Processes of Americanization in British children's use of the Internet," *Area*, 33: 153–60.

Hongladarom, S. (2011) "Personal identity and the self in the online and offline world," *Minds and Machines*, 21: 533–48.

Huang, S. and Hsu, C.H.C. (2009) "Travel motivation: Linking theory to practice," *International Journal of Culture, Tourism and Hospitality Research*, 3: 287–95.

Huh W-K. (2006) "A geography of virtual universities in Korea." Paper presented at the Annual Meeting of the IGU Commission on the Geography of the Information Society, Sydney.

Hume, D. (2008) "Of personal identity," in J. Perry (ed.) *Personal Identity*, Berkeley: University of California Press, 161–72.

Hwang, W., Jung H.-S., and Salvendy, G. (2006) "Internationalisation of e-commerce: A comparison of online shopping preferences among Korean, Turkish and US populations," *Behaviour and Information Technology*, 25: 3–18.

Hyland, K. (2011) "The presentation of self in scholarly life: Identity and marginalization in academic homepages," *English for Specific Purposes*, 30: 286–97.

IAB Australia (2012) "Australian online landscape review." Online. Available at: www.aimia.com.au/enews/iab/Website/Nielsen%20Online%20Landscape%20Review%20June%202012%20Media%20Pack.pdf (accessed June 4, 2013).

Ilchman, W.F. (1970) "New time in old clocks: productivity, development and comparative public administration," in D. Waldo (ed.) *Temporal Dimensions of Development Administration*, Durham, NC: Duke University Press, 135–78.

Ingham, J., Purvis, M. and Clarke, D.B. (1999) "Hearing places, making spaces: Sonorous geographies, ephemeral rhythms, and the Blackburn warehouse parties. Environment and Planning D," *Society and Space*, 17: 283–305.

Internet World Stats (2013) "Facebook users in the world." Online. Available at: www.internetworldstats.com/facebook.htm (accessed June 20, 2013).

Ipsos (2012a) "Interconnected world: Shopping and personal finance." Online. Available at: www.ipsos-na.com/download/pr.aspx?id=11513 (accessed June 2, 2013).

Ipsos (2012b) "Ipsos in China." Online. Available at: www.ipsos.com.cn/en/node/1983 (accessed June 2, 2013).

Iso-Ahola, S.E. (1982) "Toward a social psychological theory of tourism motivation: A rejoinder," *Annals of Tourism Research*, 9: 256–62.

ITU (International Telecommunication Union) (2003) "The birth of broadband." Online. Available at: www.itu.int/osg/spu/publications/birthofbroadband/faq.html (accessed November 27, 2013).

ITU (International Telecommunication Union) (2010) "Statistics." Online. Available at: www.itu.int/ITU-D/ict/statistics/material/graphs/movile_reg-09.jpg (accessed November 27, 2013).

ITU (International Telecommunication Union) (2011) "The world in 2011: ICT facts and figures." Online. Available at: www.itu.int/ITU-D/ict/facts/2011/material/ICTFactsFigures2011.pdf (accessed November 27, 2013).

ITU (International Telecommunication Union) (2012) "Statistical highlights 2012." Online. Available at: www.itu.int/ITU-D/ict/statistics/ (accessed September 5, 2012).

ITU (International Telecommunication Union) (2013) "Statistics." Online. Available at: www.itu.int/en/ITU-D/Statistics/Pages/stat/default.aspx (accessed April 22, 2013).

Jackson, P. (1980) "A plea for cultural geography," *Area*, 12: 110–13.

Jakle, J.A., Brunn, S.D. and Roseman, C.C. (1976) *Human Spatial Behavior*, New York: Duxbury.

James, J. (2003) "Sustainable Internet access for the rural poor? Elements of an emerging Indian model," *Futures*, 35: 461–72.

James, W. (1890) *Principles of Psychology*, New York: Holt.

Janelle, D.G. (1973) "Measuring human extensibility in a shrinking world," *The Journal of Geography*, 72: 8–15.

Järv, O., Ahas, R., Saluveer, E., Derruder, B. and Witlux, F. (2012) "Mobile phones in a traffic flow: A geographical perspective to evening rush hour traffic analysis using call detail records," *PLOS One*, 7: 1–11.

Järv, O., Ahas, R. and Witlux, F. (2014) "Understanding monthly variability in human activity spaces: A twelve-month study using mobile phone call detail records," *Transportation Research C*, 38: 122–35.

JiWire (2010) "Global Wi-Fi finder." Online. Available at: http://v4.jiwire.com/search-hotspot-locations.htm (accessed November 27, 2013).

Jones, B.W., Spiegel, B. and Malecki, E.J. (2010) "Blog links as pipelines to buzz elsewhere: The case of New York theater blogs," *Environment and Planning B: Planning and Design*, 37: 99–111.

Jordan, B. (2009) "Introduction: Blurring boundaries: The 'real' and the 'virtual' in hybrid spaces," *Human Organization*, 68: 181–93.

Kaminer, N. and Braunstein, Y.M. (1998) "Bibliometric analysis of the impact of Internet use on scholarly productivity," *Journal of the American Society for Information Science*, 49: 720–30.

Kanngieser, A. (2012) "A sonic geography of voice: Towards an affective politics," *Progress in Human Geography*, 36: 336–53.

KAPOW! (2005) "Top-ten business users of SMS text messaging." Online. Available at: www.kapow.co.uk (accessed November 27, 2013).

Kauffman, R. and Techatassanasoontorn, A.A. (2009) "Understanding early diffusion of digital wireless phones," *Telecommunications Policy*, 33, 432–50.

Kaufmann, V. (2002) *Re-thinking Mobility: Contemporary Sociology*, Aldershot: Ashgate.

Keizer, G. (2010) *The Unwanted Sound of Everything We Want: A Book about Noise*, New York: Public Affairs.

Kellerman, A. (1984) "Telecommunications and the geography of metropolitan areas," *Progress in Human Geography*, 8: 222–46.

Kellerman, A. (1989) *Time, Space, and Society: Geographical Societal Perspectives*, Dordrecht: Kluwer.

Kellerman, A. (1993) *Telecommunications and Geography*, London and New York: Belhaven (Wiley).

Kellerman, A. (1999) "Leading nations in the adoption of communications media 1975–1995," *Urban Geography*, 20: 377–89.

Kellerman, A. (2002) *The Internet on Earth: A Geography of Information*, London: Wiley.

Kellerman, A. (2006a) *Personal Mobilities*, London and New York: Routledge.

Kellerman, A. (2006b) "Broadband Penetration and its Implications: The Case of France," *Netcom*, 20: 237–46.

Kellerman, A. (2007) "Cyberspace classification and cognition: Information and communications cyberspaces," *Journal of Urban Technology*, 14: 5–32.

Kellerman, A. (2010) "Mobile broadband services and the availability of instant access to cyberspace," *Environment and Planning A*, 42: 2990–3005.

Kellerman, A. (2012a) *Daily Spatial Mobilities: Physical and Virtual*, Farnham and Burlington VT: Ashgate.

Kellerman, A. (2012b) "Potential mobilities," *Mobilities*, 7: 171–83.

Kellerman, A. (2013) "The satisfaction of human needs in physical and virtual spaces," *The Professional Geographer*, DOI: 10.1080/00330124.2013.848760.

Kellerman, A. and Paradiso, M. (2007) "Geographical location in the information age: From destiny to opportunity?," *GeoJournal*, 70: 195–211.

Kennedy, H. (2006) "Beyond anonymity, or future directions for Internet identity research," *New Media and Society*, 8: 859–76.

Kenrick, D.T., Griskevicius, V., Neuberg, S. and Schaller, M. (2010) "Renovating the pyramid of needs: Contemporary extensions built upon ancient foundations," *Perspectives on Psychological Science*, 5: 292–314.

Kinsley, S. (2013) "The matter of 'virtual' geographies," *Progress in Human Geography*, DOI: 10.1177/0309132513506270.

Kirsch, S. (1995) "The incredible shrinking world? Technology and the production of space," *Environment and Planning D: Society and Space*, 13: 529–55.

KISA (Korea Communications Commission (2012) "2011 Survey on Internet usage."

Online. Available at: http://isis.kisa.or.kr/eng/board/?pageId=040100&bbsId=10&item Id=317 (accessed June 5, 2013).

Kitchin, R. (1998) *Cyberspace: The World in the Wires*, Chichester: Wiley.

Kitchin, R. and Dodge, M. (2011) *Code/Space*, Cambridge, MA and London: MIT Press.

Klein, A. (2012) "Slipping racism into the mainstream: A theory of information laundering," *Communication Theory*, 22: 427–48.

Kline, D. (2013) *Technologies of Choice? ICTs, Development, and the Capabilities Approach*, Cambridge, MA: MIT Press.

Knight, J. (2006) *Higher Education Crossing Borders: A Guide to the Implications of the General Agreement on Trade in Services (GATS) for Cross-border Education*, Vancouver and Paris: Commonwealth of Learning.

Knorr-Cetina, K. and Bruegger, U. (2002) "Global microstructures: The virtual societies of financial markets," *American Journal of Sociology*, 107: 905–50.

Kolowish, S. (2009) "Libraries in the future," *Inside Higher Ed.* Online. Available at: www.insiderhighered.com/print/news/2009/09/24/libraries?width=775&height=500&iframe=true (accessed April 1, 2012).

Kong, L. (2001) "Religion and technology: Refiguring place, space, identity and community," *Area*, 33: 404–13.

Kopomaa, T. (2000) *The City in Your Pocket: Birth of the Mobile Information Society*, Helsinki: Gaudeamus.

Kraemer-Mbula, Tang P. and Rush, H. (2013) "The cybercrime ecosystem: Online innovation in the shadows?," *Technological Forecasting and Social Change*, 80: 541–55.

Kreitner, R. and Kinicki, A. (2008) *Organizational Behavior*, 8th ed., New York: McGraw-Hill.

Kshetri, N.B. (2001) "Determinants of the locus of global e-commerce," *Electronic Markets*, 11: 251–7.

Kwan, M.-P. (2001) "Cyberspatial cognition and individual access to information: The behavioral foundation of cybergeography," *Environment and Planning B*, 28: 21–37.

LaBelle, B. (2010) *Acoustic Territories: Sound Culture and Everyday Life*, New York and London: Continuum.

Lacohée, H., Wakeford, N. and Pearson, I. (2003) "A social history of the mobile telephone with a view of its future," *BT Technology Journal*, 21: 203–11.

Lagerkvist, J. (2008) "Internet ideotainment in the PRC: National responses to cultural globalization," *Journal of Contemporary China*, 17: 121–40.

Lampel, J. and Bhalla, A. (2007) "The role of status seeking in online communities: Giving the gift of experience," *Journal of Computer-Mediated Communication*, 12: article 5. Online. Available at: http://jcmc.indiana.edu/vol. 12/issue2/lampel.html (accessed November 6, 2012).

Larsen, J., Axhausen, K.W. and Urry, J. (2006) "Geographies of social networks: Meetings, travel and communications," *Mobilities*, 1: 261–83.

Layne, K. and Lee, J. (2001) "Developing fully functional e-government: A four-stage model," *Government Information Quarterly*, 18: 122–36.

Lee, J. and Lee, H. (2010) "The computer-mediated communication network: Exploring the linkage between the online community and social capital," *New Media and Society*, 12: 711–27.

Lefebvre, H. (1991) *The Production of Space*. D. Nicholson-Smith (trans.) Oxford: Basil Blackwell.

Lessig, L. (2001) *The Future of Ideas: The Fate of the Commons in a Connected World*, New York: Random House.

Li, F., Whalley, J. and Williams, H. (2001) "Between physical and electronic spaces: The implications for organizations in the networked economy," *Environment and Planning A*, 33: 699–716.

Li, F., Papagiannidis, S. and Bourlakis, M. (2010) "Living in 'multiple spaces': Extending our socioeconomic environment through virtual worlds," *Environment and Planning D: Society and Space*, 28: 425–46.

Licoppe, C. (2004) "'Connected' presence: The emergence of a new repertoire for managing social relationships in a changing communication technoscape," *Environment and Planning D: Society and Space*, 22: 135–56.

Liff, A.P. (2012) "Cyberwar: A new 'absolute weapon'? The proliferation of cyberwarfare capabilities and interstate war," *The Journal of Strategic Studies*, 35: 401–28.

Lin, W-Y., Zhang, X., Jung, J.-Y. and Kim, Y.-C. (2013) "From the wired to wireless generation? Investigating teens' Internet use through the mobile phone," *Telecommunications Policy*, 37: 651–61.

Locke, J. (2008) "Of identity and diversity," in J. Perry (ed.) *Personal Identity*, Berkeley: University of California Press, 33–52.

Loewenstein, G. (2002) "Psychology of curiosity," *International Encyclopedia of the Social and Behavioral Sciences*. Online. Available at: www.sciencedirect.com (accessed April 1, 2012).

Loo, B.P.Y. (2012) *The E-Society*, New York: Nova Science Publishers.

Lowenthal, D. (1961) "Geography, experience and imagination: Towards a geographical epistemology," *Annals of the Association of American Geographers*, 51: 241–60.

Luca, A.M. (2000) "Positive self-assessment at work: Positive illusions, work motivation, and career esteem," unpublished doctoral dissertation. University of California Los Angeles (UCLA).

Mabe, M. (2003) "The growth and number of journals," *Serials*, 16: 191–7.

Mabe, M. and Amin, M. (2002) "Dr Jekyll and Dr Hyde: Author-reader asymmetries in scholarly publishing," *Aslib Proceedings: New Information Perspectives*, 54: 149–57.

McLuhan, M. (1964) *Understanding Media: The Extensions of Man*, New York: Macmillan.

Madden, M., Lenhart, A. and Duggan, M. (2013) *Teens and technology 2013*. PewInternet. Online. Available at: www.pewinternet.org/Reports/2013/Teens-and-Tech.aspx (accessed October 22, 2013).

Malecki, E.J. (2002) "The economic geography of the Internet's infrastructure," *Economic Geography*, 78: 399–424.

Malecki, E.J. (2003) "Digital development in rural areas: Potentials and pitfalls," *Journal of Rural Studies*, 19: 201–14.

Malecki, E.J. and Moriset, B. (2008) *The Digital Economy: Business Organization, Production Processes and Regional Developments*, London and New York: Routledge.

Malecki, E.J. and Wei, H. (2009) "A wired world: The evolving geography of submarine cables and the shift to Asia," *Annals of the Association of American Geographers*, 99: 360–82.

Maslow, A.H. (1943) "A theory of human motivation," *Psychological Review*, 50: 370–96.

Maslow, A.H. (1954) *Motivation and Personality*, New York: Harper.

Maslow, A.H. (1968) *Toward a Psychology of Being*, New York: Van Nostrand.

Massey, D. (1992) "Politics and space/time," *New Left Review*, 196: 65–84.

Matless, D. (2005) "Sonic geography in a nature region," *Social and Cultural Geography*, 6: 745–66.

Matusitz, J. and Breen G.-M. (2007) "E-health: A new kind of E-health," *Social Work in Public Health*, 23: 95–113.

Meishar-Tal, H. (2006) "The Internet and social dynamics," unpublished doctoral dissertation. Department of Geography, University of Haifa (Hebrew).

Merisalo, M., Makkonen, T. and Inkinen, T. (2013) "Creative and knowledge-intensive teleworkers' relation to e-capital in the Helsinki metropolitan area," *International Journal of Knowledge-Based Development*, 4: 204–20.

Merriam-Webster Dictionary (2013) Presentism. Online. Available at: www.merriam-webster.com/dictionary/presentism (accessed July 15, 2013).

Merrifield, A. (1993) "Place and space: A Lefebvrian reconciliation," *Transactions of the British Institute of Geographers*, 18: 516–31.

Meyer, M. (1999) "Demand management as an element of transportation policy: Using carrots and sticks to influence travel behavior," *Transportation Research A*, 33: 575–99.

Miller, H. and Mather, R. (1998) "The presentation of self in WWW home pages." Paper presented at IRISS 1998, Bristol. Online. Available at: www.intute.ac.uk/socialsciences/archive/iriss/papers/paper21.htm (accessed May 13, 2013).

Mills, N. and Whitacre, B. (2003) "Understanding the metropolitan-non-metropolitan divide," *Growth and Change*, 34: 219–43.

Mills, K. (2002) "Cybernations: Identity, self-determination, democracy and the 'Internet effect' in the emerging information order," *Global Society*, 16: 69–87.

Mitchell, W.J. (1995) *City of Bits: Space, Place, and the Infobahn*, Cambridge, MA: MIT Press.

Mok, D., Wellman, B. and Carrasco, J. (2010) "Does distance matter in the age of the Internet?," *Urban Studies*, 47: 2747–83.

Mokhtarian, P.L. (2000) "Telecommunications and travel, in *Transportation in the new millennium*." Washington, D.C.: Transportation Research Board. Online. Available at: www4.nationalacademies.org/trb/homepage.nsf/web/millenium_papers (accessed April 1, 2012).

Moriset, B. and Malecki, E.J. (2009) "Organization versus space: The paradoxical geographies of the digital economy," *Geography Compass*, 3: 256–74.

Morozov, E. (2011) *The Net Delusion: The Dark Side of Internet Freedom*, New York: Public Affairs.

Mossberger, K., Tolbert, C.J. and Franko, W.W. (2013) *Digital Cities: The Internet and the Geography of Opportunity*, New York: Oxford University Press.

Murakami Wood, D. and Graham, S. (2006) "Permeable boundaries in the software-sorted society: Surveillance and the differentiation of mobility," in M. Sheller and J. Urry (eds.) *Mobile Technologies and the City*, London: Routledge, 177–91.

Myers, D.G. (2009) *Psychology in Everyday Life*, New York: Worth.

Nancy, J.L. (2007) *Listening*, C. Mandell (trans.) New York: Fordham University Press.

NCJRS (National Criminal Justice Reference Service) (2013) "Identity theft – facts and figures." Online. Available at: www.ncjrs.gov/spotlight/identity_theft/facts.html (accessed July 10, 2013).

NSF (National Science Foundation) (2008) *Science and Engineering Indicators*. Online. Available at: www.nsf.gov/statistics/seind06/ (accessed April 1, 2012).

Negroponte, N. (1995) *Being Digital*, New York: Alfred A. Knopf.

Neher, A. (1991) "Maslow's theory of motivation: A critique," *Journal of Humanistic Psychology*, 31: 89–112.

Neilan, C. (2010) "Academic publishers seeing strong growth from e-book sales."

*TheBookseller.com*. Online. Available at: www.thebookseller.com/news/academic-publishers-seeing-strong-growth-e-book-sales.html (accessed April 1, 2012).

Nielsen (2012) "Understanding growth markets: India and China." Online. Available at: www.tnhindia.in/statistics_kit/statistics.pdf (accessed June 4, 2013).

Nowotny, H. (2008) *Insatiable Curiosity: Innovation in a Fragile Future*, M. Cohen (trans.) Cambridge, MA: MIT Press.

NTIA (National Telecommunications and Information Administration) (2002) "A nation online: How Americans are expanding their use of the Internet." Online. Available at: www.ntia.doc.gov/ntiahome/dn/anationonline2.pdf (accessed May 30, 2013).

OECD (Organization for Economic Cooperation and Development) (2012a) "Government at a glance 2011." Online. Available at: www.oecd-ilibrary.org/sites/gov_glance-2011-en/11/02/index.html?contentType=/ns/Chapter,/ns/StatisticalPublication&itemId=/content/chapter/gov_glance-2011-55-en&containerItemId=/content/serial/22214399&accessItemIds=&mimeType=text/html (accessed November 28, 2013).

OECD (Organization for Economic Cooperation and Development) (2012b) "Public sector innovation and e-government." Online. Available at: www.oecd.org/gov/public-innovation/ (accessed November 28, 2013).

Ogden, M.R. (1994) "Politics in a parallel universe," *Futures*, 26: 713–29.

O'Kelly, M.E. and Grubesic, T.H. (2002) "Backbone topology, access and the commercial Internet, 1997–2000," *Environment and Planning B*, 29: 533–52.

Oldenburg, R. (2000) *Celebrating the Third Place: Inspiring Stories about the 'Great Good Places' at the Heart of Our Communities*, New York: Marlowe & Co.

Ong, W.J. (1971) *Rhetoric, Romance, and Technology*, Ithaca: Cornell University Press.

Oremus, W. (2013) "A map of the global Internet so pretty you can hang it on your wall," *Slate*. Online. Available at: www.slate.com/blogs/future-tense/2013/02/01/telegeography_s_gorgeous_map_of_the_global_internet (accessed April 14, 2013).

Pal, S.K., Pandey, G.S., Kesari, A., Choudhuri, G. and Mittal, B. (2002) "E-health: E-health and hospital of the future," *Journal of Scientific and Industrial Research*, 61: 414–22.

Papacharissi, Z. (2002) "The presentation of self in virtual life: Characteristics of personal home pages," *Journalism and Mass Communication Quarterly*, 79: 643–60.

Paradiso, M. (2011) "Google and the Internet: A megaproject nesting within another megaproject," in S. Brunn (ed.) *Engineering Earth: The Impacts of Megaengineering Projects*, Dordrecht: Kluwer, Vol. 1: 49–65.

Park, H. (2001) "Cultural impact on Internet connectivity and its implication," *Journal of Euromarketing*, 10: 5–22.

Pasternak, C. (2007) "Curiosity and quest," in C. Pasternak (ed.) *What Makes Us Human?*, Oxford: Oneworld, 114–32.

Paterson, M. (2006) "Feel the presence: Technologies of touch and distance," *Environment and Planning D: Society and Space*, 24: 691–708.

Pearce, P.L. (1993) "Fundamentals of tourist motivations," in D.G. Pearce and R.W. Butler (eds.) *Tourism Research: Critiques and Challenges*, London and New York: Routledge, 113–34.

Pearsall, J. (2002) *Concise Oxford English Dictionary*, Oxford: Oxford University Press.

Péruch, P., Gaunet, F., Thinus-Blanc, C. and Loomis, J. (2000) "Understanding and learning virtual spaces," in R. Kitchin and S. Freundschuh (eds.) *Cognitive Mapping: Past, Present, and Future*, London: Routledge, 108–15.

PewInternet (2012a) "How Americans use their cell phones – specific activities." Online. Available at: http://pewinternet.org/Commentary/2012/February/Pew-Internet-Mobile.aspx (accessed June 2, 2013).

PewInternet (2012b) "Mobile health 2012." Online. Available at: www.pewinternet.org/Reports/2012/Mobile-Health.aspx (accessed November 2, 2013).

PewInternet (2013a) "Trend data (adults)." Online. Available at: http://pewinternet.org/Trend-Data-(Adults)/Online-Activities-Total.aspx (accessed May 27, 2013).

PewInternet (2013b) "Pew Internet: Social networking (full detail)." Online. Available at: http://pewinternet.org/Commentary/2012/March/Pew-Internet-social-networking-full-detail.aspx# (accessed June 30, 2013).

PewInternet (2013c) "Trend data (adults)." Online. Available at: http://pewinternet.org/Trend-Data-(Adults)/Whos-Online.aspx (accessed October 21, 2013).

PewInternet (2013d) "Health online 2013." Online. Available at: http://pewinternet.org/Reports/2013/Health-online.aspx (accessed November 2, 2013).

Phillips, R. (2010) "The impact agenda and geographies of curiosity," *Transactions of the Institute of British Geographers*, 35: 447–52.

Pickerill, J. (2008) "Open access publishing: Hypocrisy and confusion in geography," *Antipode*, 40: 719–23.

Pinkerton, A., Young, S. and Dodds, K. (2011) "Weapons of mass communication: The securitization of social networking sites," *Political Geography*, 30: 115–17.

Polzer-Debruyne, A.M. (2008) "Psychological and workplace attributes that influence personal web use (PWU)," unpublished doctoral dissertation. Albany NZ: Massey University.

Porter, L.V. and Sallot, L.M. (2003) "The Internet and public relations: Investigating practitioners roles and World Wide Web use," *Journalism and Mass Communication Quarterly*, 80: 603–22.

Pósfai, M. and Féjer, A. (2008) "The eHungary programme 2.0," *Innovation*, 21: 407–15.

Rainie, L. and Wellman, B. (2012) *Networked: The New Social Operating System*, Cambridge, MA: MIT Press.

Red, F. and Kwan, M-P. (2009) "The impact of the Internet on human activity-travel patterns: Analysis of gender differences using multi-group structural equation models," *Journal of Transport Geography*, 17: 440–50.

Reid, T. (2008) "Of identity," in J. Perry (ed.) *Personal Identity*, Berkeley: University of California Press, 107–12.

Renear, A. and Palmer, C. (2009) "Strategic reading, ontologies, and the future of scientific publishing," *Science*, 325: 828–35.

Reporters without Borders (RWB) (2012) "2011–2012 Press freedom index: How the index is compiled." Online. Available at: http://en.rsf.org/press-freedom-index-2011–2012,1043.html (accessed July 8, 2013).

Rheingold, H. (1993) "A slice of life in my virtual community," in L.M. Harasim (ed.) *Global Networks: Computers and International Communication*, Cambridge, MA: MIT Press, 57–82.

Rodaway, P. (1994) *Sensuous Geographies: Body, Sense and Place*, London and New York: Routledge.

Rogers, E.M. (1995) *Diffusion of Innovations*, 4th ed., New York: The Free Press.

Royal Pingdom (2013) "The top 100 web hosting countries." Online. Available at: http://royal.pingdom.com/2013/03/14web-hosting-countries-2013/ (accessed April 14, 2013).

Rushkoff, D. (2013) *Present Shock: When Everything Happens Now*, New York: Penguin.

Sack, R. (1980) *Conceptions of Space in Social Thought*, London: Macmillan.

Sakari, T. (2013) "The use of e-government services and the Internet: The role of socio-demographic' economic and geographical predictors," *Telecommunications Policy*, 37: 413–22.

Sauer, C.O. (1941) "The personality of Mexico," *Geographical Review*, 31: 353–64.

Schafer, M. (1977) "The soundscape," in C. Kelly (eds.) *Whitechapel Gallery London*, Cambridge, MA: MIT Press, 110–112.

Schafer, M. (1994) *The Soundscape: Our Sonic Environment and the Turning of the World*, Rochester, VT: Destiny Books.

Scherr Technology (2009) "Internet, mobile, broadband, and social media world usage statistics 2009." Online. Available at: www.scherrtech.com/wordpress/2009/05/16/internet-mobile-broadband-social-media-usage-statistics-2009/ (accessed August 1, 2010).

Schmidt, E. and Cohen, J. (2013) *The New Digital Age: Reshaping the Future of People, Nations and Business*, New York: Alfred A. Knopf.

Schrag, Z.M. (1994) "Navigating cyberspace – maps and agents: Different uses of computer networks call for different interfaces," in G.C. Staple (ed.) *Telegeography 1994: Global Telecommunications Traffic*, Washington, DC: Telegeography, Inc., 44–52.

Schwanen, T. and Kwan, M-P. (2008) "The Internet, mobile phone and space-time constraints," *Geoforum*, 39: 1362–77.

Schwanen, T., Dijst, M. and Kwan, M-P. (2008) "ICTs and the decoupling of everyday activities, space and time: Introduction," *Tijdschrift voor Economische en Sociale Geografie*, 99: 519–27.

Shamai, S., and Kellerman, A. (1985) "Conceptual and experimental aspects of regional awareness: an Israeli case study," *Tijdschrift voor Economische en Sociale Geografie*, 76: 88–99.

Shaw, I.G.R. and Warf, B. (2009) "Worlds of affect: Virtual geographies of video games," *Environment and Planning A*, 41: 1332–43.

Sheller, M. (2004) "Mobile publics: Beyond the network perspective," *Environment and Planning D: Society and Space*, 22, 39–52.

Shields, R. (2003) *The Virtual*, London and New York: Routledge.

Shiu, E.C.C. and Dawson, J.A. (2004) "Comparing the impacts of Internet technology and national culture on online usage and purchase from a four-country perspective," *Journal of Retailing and Consumer Services*, 11: 385–94.

Shoemaker, S. (2008) "Personal identity and memory," in J. Perry (ed.) *Personal Identity*, Berkeley: University of California Press, 119–34.

Shum, S. (1990) "Real and virtual spaces: Mapping from spatial cognition to hypertext," *Hypermedia*, 2, 133–58.

Sim, S. (2007) *Manifesto for Silence: Confronting the Politics and Culture of Noise*, Edinburgh: Edinburgh University Press.

Simonsen, K. (1996) "What kind of space in what kind of social theory?," *Progress in Human Geography*, 20: 494–512.

Smith, R. (2004) "Access to healthcare via telehealth: Experiences from the Pacific," *Perspectives on Global Development and Technology*, 3: 197–211.

Soja, E.W. (1989) *Postmodern Geographies: The Reassertion of Space in Critical Social Theory*, London: Verso.

Soja, E.W. (1996) *Thirdspace: Journeys to Los Angeles and Other Real-and-Imagined Places*, Cambridge, MA: Blackwell.

Staats, S., Panek P.E. and Cosmar, D. (2006) "Predicting travel attitudes among university faculty after 9/11," *The Journal of Psychology*, 140: 121–32.

*Stanford Encyclopedia of Philosophy* (2013) "Personal identity." Online. Available HTTP: http://plato.stanford.edu/entries/identity-personal/ (accessed May 9, 2013).

Stat Owl (2013a) "Operating systems market share." Online. Available at: www.statowl.com/operating_system_market_share.php (accessed April 18, 2013).

Stat Owl (2013b) "Web browser market share." Online. Available at: http://statowl.com/web_browser_market_share.php (accessed April 18, 2013).

Stat Owl (2013c) "Search engine market share." Online. Available at: www.statowl.com/search_engine_market_share.php (accessed April 18, 2013).

Statistics Canada (2004) "E-commerce: Household shopping on the Internet." www.statcan.ca/Daily/English/040923/d040923a.htm [no longer online], (accessed May 29, 2013).

Stewart, D.O. (2011) "Online gambling five years after UIGEA." American Gaming Association (AGA). Online. Available at: www.amricangaming.org/sites/default/files/uploads/docs/final_online_gambling_white_paper_5–18–11.pdf (accessed November 11, 2013).

Swyngedouw, E.A. (1992) "Territorial organization and the space/technology nexus," *Transactions of the British Institute of Geographers*, 17: 417–33.

Swyngedouw, E.A. (1993) "Communication, mobility and the struggle for power over space," in G. Giannopoulos and A. Gillespie (eds.) *Transport and Communications in the New Europe*, London: Belhaven, 305–25.

Symantec (2010) "Global Internet security threat report: Trends for 2009." Online. Available at: http://eval.symantec.com/mktginfo/enterprise/white_papers/b-whitepaper_internet_security_threat_report_xv_04–2010.en-us.pdf (accessed July 9, 2013).

Symantec (2012) "The 2011 threat landscape." Online. Available at: www.symantec.com/threatreport/ (accessed November 5, 2012).

Tang, J., Zhang, D. and Yao, L. (2007) "Social network extraction of academic researchers," in Seventh IEEE International Conference on Data Mining. Online. Available at: http://keg.cs.tsinghua.edu.cn/../ICDM07-Tang-et-al-Academic-Network-Extraction.pdf (accessed May 21, 2013).

Tapscott, D. and Williams, A.D. (2010) *Wikinomics: How Mass Collaboration Changes Everything*, New York: Portfolio/Penguin.

Tay, L. and Diener, E. (2011) "Needs and subjective well-being around the world," *Journal of Personality and Social Psychology*, 101: 354–65.

Telegeography (2013a) "Submarine cable map." Online. Available at: www.submarine-cablemap.com (accessed April 14, 2013).

Telegeography (2013b) "Internet exchange map." Online. Available at: www.telegeography.com/telecom-resources/submarine-cable-map/index.html (accessed April 14, 2013).

Telework Association (2012) "How many people telework in the UK?" Online. Available at: www.tca.org.uk/content/how-many-people-telework-uk (accessed May 28, 2013).

Tenpir, C., Read, E., Manoff, M., Baker, G., Nicholas, D. and King, D.W. (2007) "What does usage data tell us about our users?" Presentation at Online Information Conference London, in H. Jezzard (ed.) *Online Information 2007*, London: Incisive Media, 80–6.

*The Blog Herald* (2005) Online. Available at: www.blogherald.com/2005/10/10/the-blog-herald-blog-count-october-2005/ (accessed June 19, 2013).

Thrift, N. (1995) "A hyperactive world," in R.J. Johnston, P.J. Taylor and M.J. Watts (eds.) *Geographies of Global Change: Remapping the World in the Late Twentieth Century*, Oxford: Blackwell, 18–35.

Tillema, T., Dijst, M. and Schwanen, T. (2010) "Decisions concerning communication modes and the influence of travel time: A situational approach," *Environment and Planning A*, 42: 2058–77.

Toffler, A. (1980) *The Third Wave*, New York: Bantam Books.

Tranos, E. (2011a) "The topology and the emerging urban geographies of the Internet backbone and aviation networks in Europe: A comparative study," *Environment and Planning A*, 43: 378–92.

Tranos, E. and Gillespie, A. (2011b) "The urban geography of Internet backbone networks in Europe: Roles and relations," *Journal of Urban Technology*, 18: 35–50.

Turkle, S. (1985) *The Second Self: Computers and the Human Spirit*, New York: Simon & Schuster.

Turkle, S. (1995) *Life on the Screen: Identity in the Age of the Internet*, New York: Simon & Schuster.

Turkle, S. (2011) *Alone Together: Why We Expect More from Technology and Less from Each Other*, New York: Basic Books.

Ullman, E. (1974) "Space and/or time: opportunity for substitution and prediction," *Transactions of the British Institute of Geographers*, 63: 135–9.

UN (United Nations) (2012) United Nations Human Right Council Resolution 20/8, 5 July 2012. Online. Available at: www.ohchr.org/EN/HRBodies/HRC/RegularSessions/Session20/Pages/ResDecStat.aspx (accessed December 7, 2013).

Urry, J. (2000) *Sociology beyond Societies: Mobilities for the Twenty-first Century*, London: Routledge.

Urry, J. (2002) "Mobility and proximity." *Sociology*, 36: 255–74.

Urry, J. (2003) *Global Complexity*, Cambridge: Polity.

US Bureau of the Census (2003) "Statistical Abstract of the United States 2002," Online. Available at: www.census.gov/prod/www/statistical_abstract.html (accessed October 30, 2013).

US Bureau of the Census (2006) "2006 Statistical Abstract: The National Data Book," Online. Available at: www.census.gov/compendia/statab/education/higher_education_institutions_and_enrollment/ (accessed August 1, 2010).

US Bureau of the Census (2010) "2010 Statistical Abstract: The National Data Book," Online. Available at: www.census.gov./compendia/statab/2010/tables/10s0593.pdf (accessed May 29, 2013).

US Bureau of the Census (2011) "2011 Statistical Abstract: The National Data Book," Table 1054. Online. Available at: www.census.gov/compendia/statab/ (accessed May 30, 2013).

US Bureau of the Census (2012a) "Statistical Abstract of the United States 2012," Tables 1055 and 1159. Online. Available at: www.census.gov/compendia/statab/2012/tables/12s1159.pdf (accessed May 30, 2013).

US Bureau of the Census (2012b) "Home-based workers in the United States: 2010." Online. Available at: www.census.gov/prod/2012pubs/p70–132pdf (accessed May 28, 2013).

Valentine, G. and Hughes, K. (2011) "Shared space, distant lives? Understanding family and intimacy at home through the lens of Internet gambling," *Transactions of the Institute of British Geographers*, 37: 242–55.

Valkenburg, P.M. and Peter, J. (2008) "Adolescents' identity experiments on the Internet: Consequences for social competence and self-concept unity," *Communication Research*, 35: 208–31.

Virilio, P. (1983) *Pure War*, New York: Semiotext(e).

Visser, E-J. and Lanzendorf, M. (2004) "Mobility and accessibility effects of B2C e-commerce: A literature review," *Tijdschrift voor Economische en Sociale Geografie*, 95: 189–205.

Voegelin, S. (2010) *Listening to Noise and Silence: Towards a Philosophy of Sound Art*, New York and London: Continuum.

Wade, C. and Tavris, C. (1987) *Psychology*, New York: Harper & Row.

Ware, M. and Mabe, M. (2009) "The stm report: An overview of scientific and scholarly journal publishing." Online. Available at: www.stm-assoc.org/industry-statistics/the-stm-report/ (accessed April 1, 2012).

Warf, B. (2013) *Global Geographies of the Internet*, Dordrecht: Springer.

Webhostin.info (2013) "Web hosting companies." Online. Available at: www.webhosting.info/webhosts/globalstats/ (accessed April 14, 2013).

Weinberger, D. (2002) *Small Pieces Loosely Joined (a unified theory of the web)*, Cambridge, MA: Perseus.

Wellman, B. (2001) "Physical place and cyberplace: The rise of personalized networking," *International Journal of Urban and Regional Research*, 25: 227–52.

Welz, C. and Wolf, F. (2010) "Incidence of Telework. Eironline: Telework in the European Union." Online. Available at: www.eurofound.europa.eu/eiro/studies/tn0910050s/tn0910050s_3.htm (accessed May 29, 2013).

Wertheim, M. (1999) *The Pearly Gates of Cyberspace: A History of Space from Dante to the Internet*, New York: W.W. Norton.

Wilkinson, E. (2011) " 'Extreme pornography' and the contested spaces of virtual citizenship," *Social and Cultural Geography*, 12: 493–508.

Wilson, M. (2003) "Chips, bits, and the law: An economic geography of Internet gambling," *Environment and Planning A*, 35: 1245–60.

Wilson, M.I., Kellerman, A. and Corey, K.E. (2013) *Global Information Society: Knowledge, Mobility and Technology*, Lanham, MD: Rowman and Littlefield.

Wieseltier, L. (1998) *Kaddish*, New York: Vintage Books.

Wood, R. and Williams, R.J. (2007) "Internet gambling: Past, present and future," in G. Smith, D. Hodgins and R.J. Williams (eds.) *Research and Measurement Issues in Gambling Studies*, San Diego, CA: Elsevier, 491–514.

WHO (World Health Organization) (2010) *E-health: Opportunities and Developments in Member States*. Online. Available at: www.who.int/goe/publications/goe_E-health_2010.pdf (accessed November 2, 2013).

Wright, J.K. (1947) "Terrae incognitae: The place of imagination in geography," *Annals of the Association of American Geographers*, 37: 1–15.

W³Techs (2013) "Usage of contents languages for websites." Online. Available at: http://w3techs.com/technologies/overview/content_language/all (accessed July 10, 2013).

Yildiz, M. (2007) "E-government research: Reviewing the literature, limitations and ways forward," *Government Information Quarterly*, 24: 646–65.

Yu, H. and Shaw, S-L. (2008) "Exploring potential human activities in physical and virtual spaces: A spatio-temporal GIS approach," *International Journal of Geographical Information Science*, 22: 409–30.

YUDU (2011) "Resources." Online. Available at: http://yudu.com/whitepapers.php (accessed April 1, 2012).

Zephoria Inc. (2013) "The top 20 valuable Facebook statistics-updated October 2013." Online. Available at: http://zephoria.com/social-media/top-15-valuable-facebook-statistics/ (accessed October 21, 2013).

Zhao, S. (2009) "Parental education and children's online health information seeking: Beyond the digital divide debate," *Social Science and Medicine*, 69: 1501–5.

Zook, M. (2003) "Underground globalization: Mapping the space of flows of the Internet adult industry," *Environment and Planning A*, 35: 1261–86.

Zook, M., Dodge, M., Aoyama, Y. and Townsend, A. (2004) "New digital geographies: Information, communication and place," in S.D. Brunn, S.L. Cutter, and J.W. Harrington (eds.) *Geography and Technology*, Dordrecht: Kluwer, 155–76.

Zook, M.A. and Graham, M. (2007a) "Mapping DigiPlace: Geocoded Internet data and the representation of place," *Environment and Planning B*, 34, 466–82.

Zook, M.A. and Graham, M. (2007b) "The creative reconstruction of the Internet: Google and the privatization of cyberspace and DigiPlace," *Geoforum*, 38: 1322–43.

Zook, M.A. and Graham, M. (2007c) "From cyberspace to DigiPlace: Visibility in an age of information and mobility," in H.J. Miller (ed.) *Societies and Cities in the Age of Instant Access*, Dordrecht: Springer, 241–54.

Zur Institute (2012) "Cybersex addiction and Internet infidelity." Online. Available at: www.zurinstitute.com/cybersex_clinicalupdate.html (accessed November 5, 2012).

# Index

Page numbers in *italics* denote tables, those in **bold** denote figures.

Printed and bound by CPI Group (UK) Ltd, Croydon, CR0 4YY

21/10/2024

01777084-0006